Purchasing Injection Molds

It's unwise to pay too much, but it's unwise to pay too little. When you pay too much you lose a little money, that is all. When you pay too little, you sometimes lose everything, because the thing you bought was incapable of doing the thing you bought it to do. The common law of business balance prohibits paying a little and getting a lot—it can't be done. If you deal with the lowest bidder, it's well to add something for the risk you run. And if you do that, you will have enough to pay for something better.

John Ruskin, 1819-1900
Author, critic, artist, British philosopher

The value of a mold is measured by the profit it creates and the quality of the product it produces.

John Von Holdt, 1919-1998
Moldmaker

Purchasing Injection Molds:
A BUYER'S GUIDE

Clare Goldsberry

THE IMM BOOK CLUB

PUBLISHED BY

ENDORSED BY

AMERICAN MOLD BUILDERS ASSOCIATION
WWW.AMBA.ORG

Injection Molding Magazine
Abby Communications Inc.
55 Madison St., Suite 770
Denver, CO 80206

tel. (303) 321-2322
fax. (303) 321-3552

First edition. ©2000 by Abby Communications Inc.
All rights reserved. No part of this book may be reproduced, disseminated, or stored in any form without written permission from the publisher.

ISBN 1-893677-07-9

Library of Congress Cataloging-in Publication Data

Goldsberry, Clare, 1947–
   Purchasing injection molds : a buyer's guide / Clare Goldsberry.—1st ed.
     p. cm.
   Includes bibliographical references and index.
   ISBN 1-893677-07-09
    1. Injection molding of plastic—Equipment and supplies—Purchasing. 2. Plastics—Molds—Purchasing. I. Title.

TP1150.G65 2000
668.4'12—dc21   00-040770

This book was designed and typeset in
Leawood and Futura by Stephen Adams, Aspen.

10 9 8 7 6 5 4 3 2 1

CONTENTS

Preface   ix

1   The Moldmaking Industry   1
Overview of the Moldmaking Industry   3
New Technology   4
Future of the Moldmaking Industry   7

2   Specifying a Mold   11
Features of a Mold   11
Considerations   14
Criteria for Selecting a Mold   15
Types of Molds   16
Runner Systems   26
How Big, How Small, How Many?   30
Will Spares Be Needed?   36
Texturing   37
Soft Tooling   37
How Much Will This Mold Cost?   40
The Foundation of Your Project   42

3   Prototype Parts and Molds   43
Stereolithography   44
High-speed Machining   44
SLS vs. Aluminum   45
Alternatives   47
Harder and Faster   48
Choice of Prototype   49

4 Requesting a Quote — 51
How Many Is Too Many? — 51
The Quoting Process — 52
Standard Practices — 54
Respecting Intellectual Property — 54
How To Get the Correct Quote — 57
Being Specific — 58
Exceptions — 64
Why the Bid Is the Bid It Is — 64
Using 3-D CAD To Estimate Mold Cost — 66
How a Quote Is Figured — 67
Understanding the Quote — 68
Offer To Negotiate — 71
The Importance of Lead Time — 71
The RFQ Says It All — 73

5 Choosing a Moldmaker — 75
The Molder in the Middle — 76
Finding the Perfect Moldmaker — 78
The TicketMaster Success Story — 80
Words of Wisdom — 85
Honesty Is the Best Policy — 86

6 Buying Molds Offshore—Is It Right for You? — 87
Where Is Offshore? — 88
The Local Advantage — 89
Offshore Mold Brokers — 91
Specifying Offshore Molds — 92
Is an Offshore Mold Right for You? — 92

7 Computers, E-commerce, and Moldmaking — 97
CAD Barriers — 98
Concurrent Engineering: Fact or Fiction? — 100
Online Bidding — 102

8 The Mold Build — 105
Delivery as a Crucial Issue — 105
Overtime — 107
Whose Fault Was That? — 108
When the Mold Is Late — 109
When Is a Mold Complete? — 109
Maintenance and Repairs — 111

Contents

9 Paying the Moldmaker   113
Paying for the Mold   114
Amortizing   116
When the Molder Buys the Mold   118

10 Guarantees and Legal Considerations   119
The Contract   120
What Guarantees To Expect   121
Mold Lien Laws   123
Alleviating Misunderstanding   124

11 A Good Relationship from the Start   127
The Mold You Need   129

Glossary   131

Bibliography   137

Index   139

PREFACE

After being in the moldmaking business for 40 years, I'm glad that finally someone has addressed the problems that the industry has with the purchasing of molds.

For years, I have been asked these questions by mold buyers: "Why do these molds cost so much?", "Why do I have such a large difference between one quote and another?", "Why does it take so long to build a mold?", and "How can I make an informed decision?"

Unfortunately, all too often the uneducated mold purchaser will make the decision based on price only—often a tragic mistake! In all my years in the moldmaking industry, I have never known a mold purchaser to receive a commendation for saving money on a mold that does not meet specifications and run properly. However, I have known of cases where an individual lost his job due to purchasing molds that do not function at all, or if they did, were unable to make parts to specification.

An educated customer is a good customer. The best molds are bought by someone who specifies the purchase and verifies the results.

So, does our industry need this book? Absolutely! In fact, I can't think of one person in our industry that would not benefit by reading this book. Our industry is going through a metamorphosis. A lot of the people who grew the moldmaking industry and have an in-depth knowledge of molds and mold processing are now retiring. Their replacements, although well schooled, do not have the benefit of many years of experience. For those people, this book will, without a doubt, be a lifesaver.

On the other hand, we have many people sitting in the corporate offices of America who don't even know what a mold looks like, making decisions on which molds to buy. Could they use this book? You betcha!

Everyone in the industry, from the apprentice on the moldmaker's shop floor to a corporate vice president of purchasing of a Fortune 100 company, could benefit from reading this book.

I commend Clare Goldsberry for her time and dedication to our industry. She has contributed much. I cannot think of a better individual to write this book than Clare. My hat goes off to her.

Bill Kushmaul, President
Tech Mold Inc., Tempe, AZ

CHAPTER 1
THE MOLDMAKING INDUSTRY

People in the moldmaking industry often say that buying a mold is a lot like buying a car.

First, molds represent as big an investment to a company as a car does to an individual, and they range from Hyundais at one end of the spectrum to Rolls Royces at the other.

Second, once you decide on a type of car, you then go from dealer to dealer in search of the best price, and the more you look the more confused you become. Prices for the same car can vary greatly from dealership to dealership. Sometimes, you're not sure if the car you thought you wanted is really the car you need—especially after you see the price! If you make the wrong decision, you're out a lot of money. You may end up with a car you do not like or one that does not meet your needs.

As with buying a car, the type of mold you buy and from whom you buy it can have long-term consequences for your business.

Most car buyers are aware of the long-term costs of making a bad decision. But those responsible for purchasing molds, either as a purchasing agent for a large original equipment manufacturer (OEM) or as an independent who needs a mold for a new product being developed, don't always understand the moldmaking industry or the process mold buying entails. When it comes to buying a mold, too many purchasers are not aware of the long-term cost implications of a bad decision.

Understanding all the costs associated with a mold is difficult. Buyers look at what they consider an insignificant plastic part and ask themselves, "How much could it possibly cost to produce this?" Well, depending on a part's configuration, size, dimensions, wall thickness, quantity, and the resin it's to be made from, the mold can be a significant expense.

Maybe you're a purchasing agent or purchasing manager of a custom or proprietary molding company. Maybe you're an original equipment manufacturer who molds products. Maybe you're a designer or engineer well-versed in the technology of moldmaking. Or, perhaps you are an inventor who knows nothing at all about plastics, moldmaking, or manufacturing who now needs to learn about them.

Whatever your position, if you deal with plastic components you will probably have some contact with moldmakers, either directly or indirectly. Knowing the moldmaking industry, how it functions, who the players are, what their role is, and what they require from you, their customer, is critical to making the relationship work successfully for everyone.

OVERVIEW OF THE MOLDMAKING INDUSTRY

Moldmaking grew hand-in-hand with injection molding, obviously, because a mold is required to produce a plastic part. Although technological advances are making it possible to get plastic parts from molds made in nontraditional ways and of nontraditional materials, the majority of the millions of plastic parts manufactured every day still come from conventional tooling made of various grades and types of steel or aluminum.

As many of these products and/or components moved—gradually at first, then more rapidly—from metal, glass, ceramic, paperboard, and other traditional materials to plastics, developments in the moldmaking industry led the way for much of the transition.

Moldmaking requires creativity—a thinking-outside-the-box mentality—in order to create new solutions using new materials for old manufacturing problems. Moldmakers tend to be creative individuals, who, as one moldmaker puts it, have the ability to "think upside down and backwards" when designing and building a mold. Many a complex problem for producing a complex plastic component has been solved by creative thinking on the part of the moldmaker.

Technology has also enhanced the moldmaking industry, making it possible for moldmakers to ply their trade in ways that are faster, more efficient, and have greater accuracy than ever before. However, the moldmaker's business today remains much like that of moldmakers 30 years ago. It remains a creative trade—a trade that involves the skill, expertise, and extensive involvement of the individual moldmaker in the process of creating a mold. And this, perhaps more than anything else, makes the moldmaking industry complex.

What do moldmakers sell? They sell their skill, their expertise, and machine time. Because not all mold shops are created equal, the price you pay for your molds may vary widely, as we'll discuss later.

NEW TECHNOLOGY

If moldmakers themselves haven't changed much over the past 30 years, their shops certainly have. Mold shops today look radically different than they did 30 years ago.

Mold shops of yesteryear contained a few milling machines, a surface grinder, and a lot of hand tools. Moldmakers were known as "handle crankers," because the machinery they used took a lot of hands-on work to accomplish the job. To remove tenths of an inch from steel required the moldmaker's steady hands on the controls, measuring, taking one more pass at the steel, and measuring again. The work was exacting and tedious.

Getting into the moldmaking business 30 years ago didn't take a lot of money, but it did take a lot of time. One mold shop, PM Mold Co. in Schaumburg, IL, was named PM because its founders, Olav Bradley and Larry Hauck, worked day jobs before going to their tiny shop at night to make molds. Today, the success of the company proves that those long hours paid off.

Machine-tool technology has changed the environment of the mold shop. Most modern mold shops have a variety of computer numerical controlled (CNC) equipment, including CNC milling machines, CNC electric discharge machines (EDM), horizontal and vertical

The Moldmaking Industry

machining centers, and—most recently—new, high-speed CNC machining centers.

All of this equipment is fitted with computer technology that has taken the handle cranking out of the process and made cutting steel not only a faster process, but a more exact and precise process as well. A machine today changes tool bits automatically as it follows pre-defined, computer-generated paths. Steel is cut through at speeds and with accuracy and precision unheard of even 20 years ago.

Complex shapes and contours that 20 years ago would have taken hundreds of hours to achieve by hand today take only a few hours using five-axis CNC milling machines.

CAD/CAM Technology

Most mold shops today have in-house mold design plus engineering capabilities, which means they are equipped with a variety of CAD/CAM (computer-aided design/computer-aided machining) equipment. Design software and computer hardware such as AutoCad, Unigraphics, and Pro/Engineer, among others, are common in mold shops today. Drawing boards and sharp pencils are a thing of the past for mold designers. CAD is used in conjunction with CAM to make the various mold components with minimal involvement from the machinist or moldmaker.

Why all this investment in expensive, computer-aided equipment? First, to reduce costs-to-manufacture. Second, to help make U.S. mold shops competitive with their offshore counterparts. Since a mold's price is determined in large part by the number of hours it takes to design, machine, and fit the many mold components, reducing those hours is critical to competitiveness.

Also, buyers of molds today expect their moldmakers to be more than just builders of molds; they also expect them to take on the challenges of project management, parts engineering, and mold design that provide for optimum manufacturing. As the responsibility of the moldmaker grows, so does the need for technology to provide these services.

Consequently, a moldmaker's services have become more technology-based as moldmaking continues to evolve from an "art" to a science.

Personnel

Changes in how mold shops operate and the types of personnel required are the result of the many technological changes in the industry. In the old days, the moldmaker was handed a set of blueprints and he built the entire mold, including all its components. A shop owner or foreman knew which moldmakers could do which type of job best, and, once given the blueprints, that moldmaker owned the job from start to finish.

Mold shops still employ journeymen moldmakers, who have all the skill and expertise necessary to build and assemble an entire mold, but this is changing. One contributor to this change is a lack of young journeymen moldmakers to replace those who are retiring. To offset this, shops employ specialty machinists who are knowledgeable and skilled in one particular area—for example, building electrodes using CNC equipment. These specialists perform their individual tasks on the mold, and the journeyman moldmaker finishes the job by doing much of the detail work, fitting, and assembly.

Other shops set up "manufacturing cells," in which one moldmaker oversees several CNC machines. He is responsible for setting up the machines and seeing that they perform their various functions flawlessly.

As CAD/CAM and machine-tool technology continue to advance, the trend is toward "unassisted machining." This means that computers in the design department continuously feed information to the machines on the shop floor. This type of lights-out production work, usually done on the graveyard shift, is allowing moldmakers to be more efficient and productive while helping them reduce the costs associated with personnel.

With customers demanding more and more involvement in new product design and total manufacturing solutions, mold shops have been forced to hire more design engineers with expertise in computers and a myriad of software programs. This shift away from machinists and moldmakers on a shop's production floor (the "back end" of the project), to more computer-oriented design and engineering people who work in the office (the "front end" of the project), can reduce the number of hours involved in building a mold and make shops more competitive.

FUTURE OF THE MOLDMAKING INDUSTRY

As in most industries in today's fast-paced manufacturing environment, consolidations have hit the moldmaking industry. Smaller shops that serve specific customers or markets, or that specialize in certain types of molds, are being acquired by larger shops seeking to gain entry into those markets or areas of expertise.

The roller-coaster economic ups and downs of moldmaking mean that some shops do not survive. In addition, the proliferation of offshore moldmaking companies in Taiwan, Hong Kong, China, Portugal, and eastern Europe has hurt the U.S. moldmaker. The competition for price and, in some cases, delivery, is too much to overcome, and certain shops are forced out of business.

Older shop owners, especially those who have no children interested in the business, often must sell—generally to a younger partner or employee—in order to retire. Some just close their doors, sell the equipment, and retire.

What all of this means is that the moldmaking industry, although still comprised of many small, family-owned shops, is taking on a new look. The successful mold shop of the future will be much larger, with the resources—both financial and technological—to provide the economies of scale needed to serve an increasingly demanding OEM customer anywhere in the world.

Joint ventures have become one way mold shops are combining talents and resources to serve customers better. For example, Dollins Tool Inc. in Independence, MO created a formal partnership with Glendan Mould Inc. in Rexdale, ON after working together informally for several years. As a result, Dollins got a badly needed 30 percent increase in moldbuilding capacity and opened a door to the Canadian market, where many of its customers are located. Glendan received similar benefits from its association with Dollins.

Another large moldmaking company, Tradesco Mold Ltd. in Toronto, combined its forces with Fairway Molds Inc. in Los Angeles and formed a new company called StackTeck Systems Inc. This arrangement meant that resources could be shared to help smooth out what is typically a very "lumpy" business.

Almost everyone in the industry agrees that the many moldmaking shops in the United States make for fierce competition, particularly among the smaller shops. For buyers, this is not all bad, as it often means lower-priced tooling. Consolidations, however, should help strengthen some of these shops and enable them to weather the down times while giving them more clout with customers.

Knowing how moldmakers operate their businesses, and what you as a mold purchaser should do to facilitate the purchasing process, will help everyone involved successfully complete a project.

CHAPTER 2
SPECIFYING A MOLD

Specifying the type of mold required to manufacture plastic parts can be challenging at best, yet this is probably the most critical component of the mold-buying process. When purchasers begin contacting moldmakers and sending out requests for quote (RFQ), they are often surprised at the many types of molds available for an application and at the many different levels of pricing.

At this point, an overall look at molds will help you understand what building a mold entails.

FEATURES OF A MOLD

A mold consists of hundreds of components, each one providing an important function in the mold's operation. Just a few of the primary components will be outlined here (Figure 2.1).

The mold base consists of metal plates that attach to a molding machine, or press, and hold the cavity ("A" half) of the mold, which produces the exterior of the part, and the core ("B" half) of the mold, which produces the interior of the part.

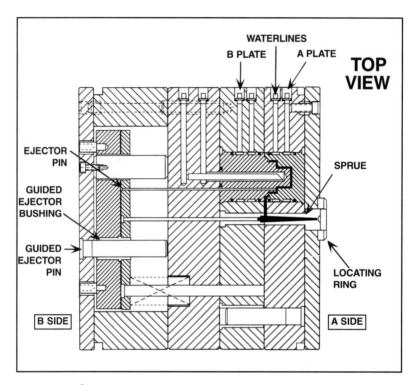

Figure 2.1 Schematic of mold showing basic components. (*Source:* Tech Mold)

The mold base also holds waterlines, or channels, through which cool water is pumped during the molding process to solidify the plastic resin. Waterlines are extremely important in the injection molding process, as cooling—or lack of—can affect molding cycle times and thus per-unit cost of parts.

A locating ring positions the mold in the molding press.

Molten plastic material is fed via a sprue into the mold's runner system.

Specifying a Mold

Ejector pins are one way to eject parts from a mold. Small, usually round, metal pins come forward from the core side as the mold opens and push the part off the core. On the inside of many plastic parts are slight round indentations left by ejector pins as the warm part was pushed from the mold. Some moldmakers will quote "guided ejection" on molds. Guided ejection mechanisms control the alignment of the ejector pins and ejector retainer plates. There are usually four in each mold, one in each quadrant.

Another method of ejecting parts is through a stripper plate. This is a moving plate that literally strips the part from the mold. Often a stripper plate is used in cases where, for cosmetic or functional reasons, ejector pin marks are unacceptable.

Other molds require slides or cams to create complex parts with undercuts, holes, external threads, or ribs.

The gate is the point at which melted plastic enters a mold. Many types of gates are used[1]:

- Pinpoint (a tiny, round gate).
- Fan (a broad fan-shaped gate).
- Flash (cousin of the fan gate).
- Sub (most widely used in high-cavitation molds).
- Sprue (used on larger parts).
- Edge (the most widely used).
- Ring (used when molding round parts).

The type of gate depends on the type of part to be molded and the part's cosmetic, manufacturing, and functional considerations. A moldmaker can help determine the optimum gate type for a specific application.

CONSIDERATIONS

When you begin the mold-buying process, one of the first things to consider is whether or not the part you want molded *can* be molded.

People unfamiliar with plastics, plastic part design, or moldmaking often think that any part can be molded of plastic, regardless of the part's dimensions or configuration. Thus, they go in search of someone who will promise to build them a mold to make exactly the part they want.

Moldmakers will usually try to build a mold to give someone the part they need, because moldmakers take pride in meeting requirements and in solving manufacturing problems. But sometimes a part just can't be molded. Or, it can be molded, but at great expense due to the design's complexity. For that reason, designing a part that lends itself to optimum molding conditions is critical to the success of the entire project.

Good part design is essential to good mold design, and both are essential to a good outcome. In fact, according to Robert Braido, health care market development manager for GW Plastics, Bethel, VT, a leading custom injection molder, 70 to 80 percent of a product's ultimate manufacturing cost is determined in the design phase.

Most designs today are computer generated using a 3-D software program. Most moldmakers can receive prints via electronic data interface (EDI), e-mail, or on a website (see Chapter 7). However, others still prefer to receive prints on paper for quoting purposes.

Specifying a Mold 15

The cost of design engineering by a moldmaker, which might include part design assistance and evaluation for manufacturability, is usually built into the cost of the mold. However, if you contract with a moldmaker for engineering services because your project isn't ready for release and you haven't put it out for bid, then you should expect to pay for those services separately.

CRITERIA FOR SELECTING A MOLD

Parts can be molded more efficiently and with less scrap and fewer quality problems if you select the optimum type of mold for your part. Different types of molds offer varying advantages, depending on what you want to achieve. From the very beginning of your search for a mold, take these factors into account:

Volume. How many parts will be required annually or per month? High-volume molds, such as large multi-cavity hot runner molds or stack molds, can be costly in terms of upfront expenditure, but cost-effective if you plan to run millions of parts annually over a period of years.

Expected service life. Molds made of aluminum or P–20 steel (P stands for prehardened) might be slightly more cost-effective upfront, but may not have the longevity of a hardened steel mold. Again, take volume into consideration, as well as the resin that will used.

Part design. The design and function of the part will dictate the kind of plastic in which the part will be molded, and the material dictates the type of mold you will use. Where will the part be manufactured—in what molding press, and under what conditions? The answers lead into manufacturability issues.

Manufacturability. Manufacturability describes the relationship between part design, mold design, and processing. Almost any part can be molded, but at what cost? Design for manufacturability (DFM) means that the part design and mold design are optimum for producing the part and achieving the best per-unit cost. The correlation among these three areas cannot be understated when choosing a mold.

Many variables determine the type of mold to be used, which is why communicating with your moldmaker during the part's design phase is essential to a good outcome. Your moldmaker can assist you in selecting the optimum mold type for your application.

TYPES OF MOLDS

As you begin evaluating types of molds that might be suitable for your application, you'll probably find you have several options—or only one or two.

Conventional or standard molds. A standard mold with an ejector-plate assembly is generally used when parting line runners are acceptable, and ejector pins, sleeves, or blades are adequate to remove the molded parts from the mold (Figure 2.2). A standard mold is sometimes referred to as a two-plate mold and consists of a cavity side and a core side.

Slide-core mold base. When an undercut or coring feature is required that cannot be formed and ejected through a

Specifying a Mold　　　　　　　　　　　　　　　　　　　　**17**

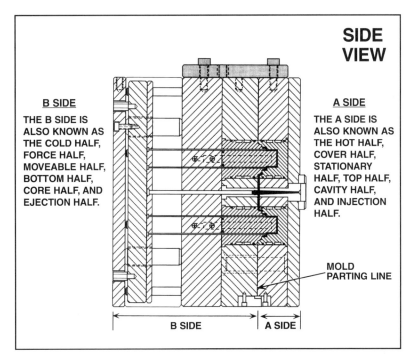

Figure 2.2 A conventional or standard mold with an ejector plate assembly is generally used when parting line runners are acceptable and ejector pins, sleeves, or blades are adequate to remove the molded plastic parts from the mold. (*Source:* Tech Mold)

standard mold opening, a slide-core mold base is generally used. The slide core is used to form the feature and is withdrawn prior to ejection of the part (Figure 2.3).

When slides are present, the mold closes, the slides come into place, the plastic is shot into the mold, the slides pull out, and the mold opens, allowing for easy removal of the part. This mold is good for parts with complex features.

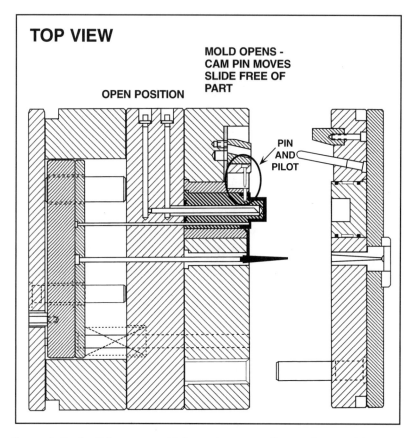

Figure 2.3 A slide-core mold base is generally used to mold a part when an undercut or coring feature is required that could not be formed and ejected through a standard mold opening. The slide core is used to form the feature and is withdrawn prior to ejection of the part. (*Source:* Tech Mold)

Stripper-plate mold. When ejector pins or blades are inadequate or objectionable in removing parts from a mold, a stripper-plate mold is generally used (Figure 2.4). Caps and closures with internal threads are examples of parts that might be molded this way.

Specifying a Mold

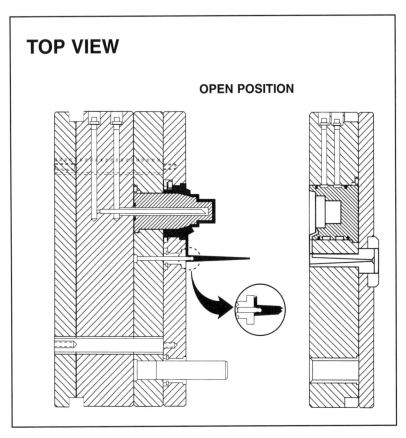

Figure 2.4 A stripper-plate mold is generally used when pins or blades are inadequate or objectionable in removing molded parts from the mold. (*Source:* Tech Mold)

Three-plate mold. When placing a gate at the side or edge of a part is objectionable because it could cause fill problems, a three-plate mold permits a central fill point that allows for uniform filling without part-weakening weldlines (Figure 2.5).

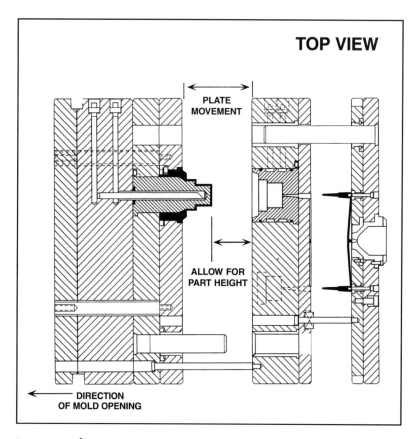

Figure 2.5 A three-plate mold is generally used when a gate position at the side or edge of the molded part would be objectionable or would likely cause no fill or extremely hard-to-fill sections in the part. When used as a central fill point in a symmetrical part, a three-plate mold allows the part to be filled uniformly without creating weldlines, which would weaken the part. (*Source:* Tech Mold)

Thread-forming (unscrewing) mold with a hot manifold. When a part requires internal threads and cannot be ejected by a stripper-plate mold without damaging the threads, a thread-forming mold generally is used.

Specifying a Mold 21

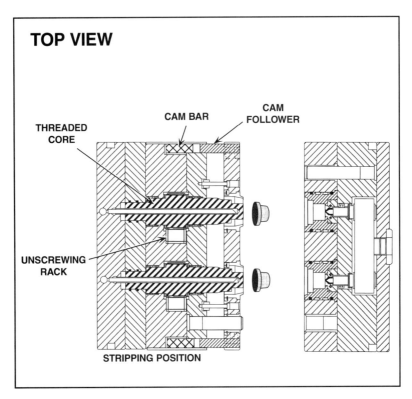

Figure 2.6 A thread-forming mold is generally used when internal threads are required in a molded part and the part cannot be ejected by a standard stripper plate mold without damaging the threads. The hot manifold is added to eliminate the runner and speed up the molding cycle. (*Source:* Tech Mold)

The hot manifold is a heating and distribution system for the resin, used for many types of molds; the system feeds from the molding machine's injection nozzle and carries the melted plastic directly to each cavity or to a secondary runner system (Figure 2.6). Thread-forming molds are used in producing plastic nuts, bolts, certain gears,

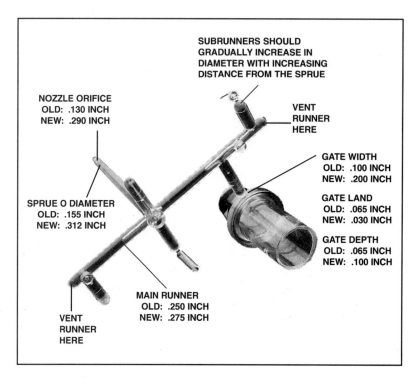

Figure 2.7 Example of a family mold, showing how balance was achieved. (*Source: Injection Molding Magazine*)

and caps and closures for the food and beverage industry. Internal threads can sometimes be stripped or molded with collapsible cores, as in soft-drink and coffee closures, respectively.

Family mold. Multiple components of a product, all made of the same resin, can be molded in a family mold cavity (Figure 2.7). A good example of a family mold's use is for a plastic model airplane or car in which parts of various sizes and shapes come still attached to one runner.

> **Family Molds Are Booming**
>
> A resurgence of family molds is taking place in Europe, according to Anne Bernhardt, president of Plastics & Computers Inc., a consulting firm in Dallas. She attributes this to a push for just-in-time manufacturing and the use of sequential gating technology for timing gate openings and closings, which results in optimum parts from a family mold. Bernhardt says family molds offer shorter production runs and more flexible manufacturing.
>
> "It's more of an advantage to mold all the dissimilar parts at the same time and not have to hang two [or more] molds and use two [or more] machines," she notes. "However, a family mold increases the complexity of the process and of the mold, requiring a more precise operation all the way around."

A common misconception is that a family mold is less expensive than separate molds when different components are required. Family molds must be carefully designed, and runner configuration is extremely critical to the molding process. Parts of varying sizes and weights must fill evenly in their respective cavities, a process called "balancing." Improper balancing means that some parts are overfilled and others are underfilled. (One solution to this problem is a patented fill-balancing technique developed by Cavity Balancing Systems of Kitchener, ON, which calls for an adjustable extra cavity called the EquaFill overflow.)

A second downside to a family mold is that cycle times can be long, which increases costs-to-manufacture. In addition, an operator is often required when a project calls for the parts to be removed from the runner system and separated. This can be done automatically, but at greater expense. Before insisting on a family mold, get an evaluation and recommendation from a moldmaker.

Multiunit die. Several sets of cores and cavities can be used in a common base in this mold, so you don't have to buy separate molds, including bases, for each part. You can purchase the cores and cavities and fit them into a standard-sized mold base. Although it is used primarily for small parts and is definitely not suitable in all cases, a multiunit die can accommodate a variety of applications. It is often a good alternative when you have a tight budget and many different components to mold.

Stack mold. Essentially, a stack mold is two molds stacked back-to-back that share a common plate. A stack mold doubles the cavitation without the need to increase press size.

Stack molds have been used extensively for small flat parts such as bottle caps and small food containers and lids. However, the technology of the stack mold is such that it can also be utilized for molding medium-sized parts such as thin-wall food containers. One company molds strawberry baskets in stack-mold systems supplied by Husky Injection Molding Systems, Bolton, ON. Even parts with complex geometry can be run in stack molds.

Running a stack mold can be more economical than running a conventional mold, in that greater productivity and efficiencies can be achieved, often without adding equipment or production space.

Interest in stack molds is high for a variety of applications in the electrical, medical, and health care products industries. Ed Cigoi, plant manager for Precise Massie, St. Petersburg, FL, commented in a magazine interview that the decision to use stack molds boils down to getting more capacity without increasing the customer's capital expenditures, thereby reducing price per part. However, he cautioned, stack molds are "obviously not for all product applications, but we offer this option to our customers when we feel the return on investment warrants it."

Because of the large size of a stack mold, support equipment such as an overhead crane is needed to handle the molds. Also, special molding expertise is required.

Although stack molds typically tend to be used for smaller applications, never underestimate the creativity of a good moldmaker. Snider Mold Co., in Mequon, WI designed and built a monster stack mold to produce covers for 500-liter water tanks for a firm in South America. The covers are about 4 ft in diameter. The aluminum mold cost about $500,000 and runs in a 1000-ton structural-foam press.[2]

Two-component molds. For molding parts that have two different materials or two different colors of material, two-component molds are becoming more popular, particularly in the consumer markets. Items such as key caps, toothbrush handles, hairbrush handles, razor handles, and various handles for tools and small appliances are utilizing two-component molding more frequently.

Two-component molds are tricky to build, and not every moldmaker has the required expertise. Usually two different material shrinkages must be considered, and good bonding between the two materials must be ensured. Also, successful molding calls for product development people with the know-how to design a two-component part suitable for molding. In addition, this type of mold takes specialized molding equipment that makes it difficult for the moldmaker to sample the mold.

You don't need to know all the details about all the types of molds that are available, but you should be familiar with a few of the options before you send out a request for quote.

Understanding molds ensures that you can get the best mold for the money, one that will meet your needs and provide optimum manufacturing in a cost-effective manner. The more you know about your own requirements, the better your moldmaker can assist you with making the right decision.

RUNNER SYSTEMS

Within a mold, two basic types of runner systems are standard.

Cold runner, or conventional. Melted plastic from a molding machine nozzle feeds through a sprue where it flows along symmetrical channels (runners) equally to all parts being filled. The plastic inside a runner is also called a runner; it solidifies at approximately the same rate as the molded part and must be removed from the mold at the end of the cycle and detached from each part (Figure 2.8).

Specifying a Mold 27

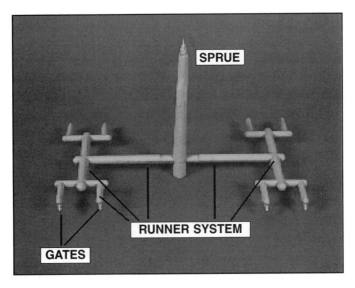

Figure 2.8 Cold runner system for a mold that produces eight parts at a time. (Source: *Injection Molding Magazine*)

The runner then is either thrown away or reground into tiny pieces for reuse where acceptable. A cold runner system often requires higher heats and/or pressures to push the material through the runner system and into the part, which can have an adverse effect on the material—depending on type—and the parts being molded. A cold runner system can also add to the cycle time required per shot.

Hot runner, or runnerless. A hot runner system is used to eliminate or lessen the use of runners that must be removed from the molded parts. The runner is heated to keep the plastic molten until it reaches the cavity. Thus, the point of entry to the mold cavity, or the gate, will have a hot drop (nozzle) suspended in the gate area.

There are various configurations of this system to accommodate gating type and part design.

The advantage of a hot runner system is that material waste is reduced and labor is lessened because there are no runners to be trimmed and thrown away or reground. This is important for components in which runner regrind is not acceptable because of quality or FDA restrictions. Hot runner systems also deliver a more consistent melt temperature to remote areas of the mold, thereby providing more acceptable molded parts and reducing scrap.

A conventional, or cold, runner system is generally less expensive than a hot runner system. However, if a part has high-volume requirements, restrictions on regrind, or needs lower per-unit pricing, a hot runner mold might be what you need to do the job (Table 2.1).

Precise Technology Inc., headquartered in Versailles, PA, specializes in molding for the consumer products industry, where the advantages of hot runner molds are critical. Precise builds and runs numerous hot runner molds.

"The biggest advantage to a hot runner mold is cost efficiency," says Bill DiMieri, director of engineering for Precise. "Every mold has a runner system—the way the hot plastic enters the mold cavity. In a hot runner system, the plastic in the runner is kept hot. Without waiting for the runner to cool, you can typically operate a hot runner mold at a faster cycle time. This is particularly true for thin-walled parts."

DiMieri also notes that eliminating regrind avoids process and contamination problems, and he considers

Specifying a Mold

Table 2.1 Hot Runner Conversion Comparison (*Source:* Mold-Masters)

	Cold runner	Hot-cold runner	Hot runner
Mold Cavitation	64	64	64
Cycle	14	13	12
Output/hour	16,457	17,723	19,200
Days/week	7	7	7
Weeks/year	50	50	50
Annual volume	138,240,000	148,873,846	161,280,000
Machine rate/hour, $	50	50	50
Cost/1000	3.04	2.82	2.60
Annual savings, $		30,000	60,000
Additional volume		10,366,846	23,040,000
Additional profit, $		21,267.69	46,080.00
Assumed profit/1000, $	2.00		
Total savings, $		51,267.69	106,080.00
Increased production, %		7.69	16.67
Total shot weight, lb	.2460	.1980	
Runner weight, lb	.0960	.0480	
Total part weight, lb	.1500	.1500	
Material/year, lb	531,360	460,578	378,000
Material reduction, %		13.32	28.86

them more "automation friendly," which translates into fewer operator steps, something the customer pays for.

Since the customer also pays for the scrap produced in the molding process, configuring the mold for a hot runner or cold runner system is critical to overall piece price.

Typically a hot runner mold is more expensive to build—estimates place the cost at anywhere from $3000 to $5000 more per cavity. However, when other savings are calculated, it may be more cost-effective to specify a hot runner mold.

One caveat about purchasing hot runner molds comes from Bob Hatch, manager of technical service and customer support for Prime Alliance in Des Moines, IA, and author of *On the Road with Bob Hatch: 100 Injection Molding Problems Solved By IMM's Troubleshooter*. Hatch recommends purchasing the entire mold from a single source to obtain a better mold. "To save time and money, most molders or tool shops will buy a hot half and fit it into their mold," says Hatch.[3] "The problem with doing it yourself is that quite often the hot half is correctly sized, but the half done in your own shop doesn't have the proper gate diameter and the land isn't sized correctly for the wall thickness of the part."

The advantages of hot runner molds are becoming more apparent in today's competitive environment, where lower costs-to-manufacture are critical for the OEM, and profit margins are critical for the molder.

HOW BIG, HOW SMALL, HOW MANY?

The number of cavities in a mold is a function of the number of parts required annually. A mold with too few cavities won't be able to meet your requirements—or will be able to meet them only if the press runs 24 hours a day, seven days a week, 52 weeks a year.

Specifying a Mold

That means no downtime for maintenance or repair on the mold, and, should the mold go down unexpectedly, it could result in a shortage of parts at a critical time.

Typically, say those in the industry, mold buyers don't define their targets with respect to manufacturing goals. Buyers should ask themselves, "Why do I want a 16-cavity mold vs. a four-cavity mold?" (Table 2.2)

The more cavities you have, the larger the mold must be to accommodate them. With a large part, more than one cavity means a mold with a footprint of several square feet, and it might be better to choose a one-cavity mold. Other smaller parts are more easily accommodated in a multicavity mold.

Volume is another consideration. Mold buyers with large-volume requirements, such as several million parts annually, often prefer to buy a large multicavity mold—one with 32 or 64 cavities, and so on. (The number of cavities must be balanced—2, 4, 8, 16, 32, or more. This is sometimes referred to as the "power of 2," meaning 2 to the second power equals 4, 2 to the third power equals 8, and so on.)

Small components like disposable medical parts or caps for pharmaceutical bottles or milk and water jugs, and other closures for food containers, are often molded in large, multicavity molds that require presses with larger tonnage.

Other mold buyers with high-volume requirements prefer several smaller, multicavity molds—perhaps four eight-cavity molds rather than one 32-cavity mold.

Table 2.2a Actual Part Production in 1 Hour
(*Source:* Tech Mold)

Cycle time	Numbers of cavities					
Seconds	1	4	8	16	24	32
4	900	3,600	7,200	14,400	21,600	28,800
4.5	800	3,200	6,400	12,800	19,200	25,600
5	720	2,880	5,760	11,520	17,280	23,040
5.5	655	2,618	5,236	10,473	15,709	20,945
6	600	2,400	4,800	9,600	14,400	19,200
6.5	554	2,215	4,431	8,862	13,292	17,723
7	514	2,057	4,114	8,229	12,343	16,457
7.5	480	1,920	3,840	7,680	11,520	15,360
8	450	1,800	3,600	7,200	10,800	14,400
8.5	424	1,694	3,388	6,776	10,165	13,553
9	400	1,600	3,200	6,400	9,600	12,800
9.5	379	1,516	3,032	6,063	9,095	12,126
10	360	1,440	2,880	5,760	8,640	11,520
10.5	343	1,371	2,743	5,486	8,229	10,971
11	327	1,309	2,618	5,236	7,855	10,473
11.5	313	1,252	2,504	5,009	7,513	10,017
12	300	1,200	2,400	4,800	7,200	9,600
12.5	288	1,152	2,304	4,608	6,912	9,216
13	277	1,108	2,215	4,431	6,646	8,862
13.5	267	1,067	2,133	4,267	6,400	8,533
14	257	1,029	2,057	4,114	6,171	8,229
14.5	248	993	1,986	3,972	5,959	7,945
15	240	960	1,920	3,840	5,760	7,680
15.5	232	929	1,858	3,716	5,574	7,432
16	225	900	1,800	3,600	5,400	7,200
16.5	218	873	1,745	3,491	5,236	6,982
17	212	847	1,694	3,388	5,082	6,776
17.5	206	823	1,646	3,291	4,937	6,583
18	200	800	1,600	3,200	4,800	6,400
18.5	195	778	1,557	3,114	4,670	6,227
20	180	720	1,440	2,880	4,320	5,760

Specifying a Mold

Table 2.2b Part Production in 1 Hour with 85% Efficiency (*Source:* Tech Mold)

Cycle time	Numbers of cavities					
Seconds	1	4	8	16	24	32
4	765	3,060	6,120	12,240	18,360	24,480
4.5	680	2,720	5,440	10,880	16,320	21,760
5	612	2,448	4,896	9,792	14,688	19,584
5.5	556	2,225	4,451	8,902	13,353	17,804
6	510	2,040	4,080	8,160	12,240	16,320
6.5	471	1,883	3,766	7,532	11,298	15,065
7	437	1,749	3,497	6,994	10,491	13,989
7.5	408	1,632	3,264	6,528	9,792	13,056
8	383	1,530	3,060	6,120	9,180	12,240
8.5	360	1,440	2,880	5,760	8,640	11,520
9	340	1,360	2,720	5,440	8,160	10,880
9.5	322	1,288	2,577	5,154	7,731	10,307
10	306	1,224	2,448	4,896	7,344	9,792
10.5	291	1,166	2,331	4,663	6,994	9,326
11	278	1,113	2,225	4,451	6,676	8,902
11.5	266	1,064	2,129	4,257	6,386	8,515
12	255	1,020	2,040	4,080	6,120	8,160
12.5	245	979	1,958	3,917	5,875	7,834
13	235	942	1,883	3,766	5,649	7,532
13.5	227	907	1,813	3,627	5,440	7,253
14	219	874	1,749	3,497	5,246	6,994
14.5	211	844	1,688	3,377	5,065	6,753
15	204	816	1,632	3,264	4,896	6,528
15.5	197	790	1,579	3,159	4,738	6,317
16	191	765	1,530	3,060	4,590	6,120
16.5	185	742	1,484	2,967	4,451	5,935
17	180	720	1,440	2,880	4,320	5,760
17.5	175	699	1,399	2,798	4,197	5,595
18	170	680	1,360	2,720	4,080	5,440
18.5	165	662	1,323	2,646	3,970	5,293
20	153	612	1,224	2,448	3,672	4,896

Table 2.3c Part Production in 1 Year with 85% Efficiency (*Source:* Tech Mold)

Cycle time Seconds	Numbers of cavities					
	1	4	8	16	24	32
4	6,701,400	26,805,600	53,611,200	107,222,400	160,833,600	214,444,800
4.5	5,956,800	23,827,200	47,654,400	95,308,800	142,963,200	190,617,600
5	5,361,120	21,444,480	42,888,960	85,777,920	128,666,880	171,555,840
5.5	4,873,745	19,494,982	38,989,964	77,979,927	116,969,891	155,959,855
6	4,467,600	17,870,400	35,740,800	71,481,600	107,222,400	142,963,200
6.5	4,123,938	16,495,754	32,991,508	65,983,015	98,974,523	131,966,031
7	3,829,371	15,317,486	30,634,971	61,269,943	91,904,914	122,539,886
7.5	3,574,080	14,296,320	28,592,640	57,185,280	85,777,920	114,370,560
8	3,350,700	13,402,800	26,805,600	53,611,200	80,416,800	107,222,400
8.5	3,153,600	12,614,400	25,228,800	50,457,600	75,686,400	100,915,200
9	2,978,400	11,913,600	23,827,200	47,654,400	71,481,600	95,308,800
9.5	2,821,642	11,286,568	22,573,137	45,146,274	67,719,411	90,292,547
10	2,680,560	10,722,240	21,444,480	42,888,960	64,333,440	85,777,920
10.5	2,552,914	10,211,657	20,423,314	40,846,629	61,269,943	81,693,257
11	2,436,873	9,747,491	19,494,982	38,989,964	58,484,945	77,979,927
11.5	2,330,922	9,323,687	18,647,374	37,294,748	55,942,122	74,589,496
12	2,233,800	8,935,200	17,870,400	35,740,800	53,611,200	71,481,600
12.5	2,144,448	8,577,792	17,155,584	34,311,168	51,466,752	68,622,336
13	2,061,969	8,247,877	16,495,754	32,991,508	49,487,262	65,983,015
13.5	1,985,600	7,942,400	15,884,800	31,769,600	47,654,400	63,539,200
14	1,914,686	7,658,743	15,317,486	30,634,971	45,952,457	61,269,943
14.5	1,848,662	7,394,648	14,789,297	29,578,593	44,367,890	59,157,186
15	1,787,040	7,148,160	14,296,320	28,592,640	42,888,960	57,185,280
15.5	1,729,394	6,917,574	13,835,148	27,670,297	41,505,445	55,340,594
16	1,675,350	6,701,400	13,402,800	26,805,600	40,208,400	53,611,200
16.5	1,624,582	6,498,327	12,996,655	25,993,309	38,989,964	51,986,618
17	1,576,800	6,307,200	12,614,400	25,228,800	37,843,200	50,457,600
17.5	1,531,749	6,126,994	12,253,989	24,507,977	36,761,966	49,015,954
18	1,489,200	5,956,800	11,913,600	23,827,200	35,740,800	47,654,400
18.5	1,448,951	5,795,805	11,591,611	23,183,222	34,774,832	46,366,443
20	1,340,280	5,361,120	10,722,240	21,444,480	32,166,720	42,888,960

The strategy of using four eight-cavity molds means that the molds have a smaller footprint, can run in smaller presses that cost less to operate, and can often reduce overall costs-to-manufacture.

The above choices reflect the two differing strategies in mold manufacturing today: A single, large, multicavity mold with 32 cavities has a much larger footprint and requires a much larger press to run. Although the piece part price is exponentially less with more cavities, the fact that the mold must run in a larger press that costs more to operate often offsets any real savings in piece part price.

Delphi Packard Electric, Clinton, MS, has a large, in-house molding facility with more than 100 presses. The company decided to reverse its philosophy and went from using large multicavity molds to smaller four- and eight-cavity molds. It found the molds much easier to move in and out of the presses, and if one mold had to go down for repair and maintenance, the other molds kept on producing.

Others point out that you rarely see a 64-cavity mold with all 64 cavities in use. The tendency is to block off a bad cavity until there's a break in the production cycle before taking the mold out for repair. It is not uncommon for a 64-cavity mold to be running with a dozen cavities blocked off. From a production standpoint, this is neither cost-effective nor efficient.

Some molds are built to run in a specific type of injection molding press and are literally "married" to the press. A few larger moldmakers can work with machinery manufacturers to accomplish this. For example,

Husky Injection Molding Systems builds molding systems comprised of the mold and the press it will run in, and the two are shipped together. This is usually for applications such as packaging, caps, and closures, or in cases where the mold is never removed from the press except for maintenance.

In making a decision as to size and number of cavities, the prudent approach is to work with a moldmaker to evaluate the numbers of parts required weighed against the cost-effectiveness of large vs. small multicavity molds.

WILL SPARES BE NEEDED?

Multicavity mold purchasers sometimes require spare cores and cavities to be included as part of the deal. Spares can minimize downtime, ensuring that if a cavity goes down for any reason, production levels can be maintained by exchanging the down core/cavity with a good one. This is where the idea of interchangeability comes into play as well, although there is a difference between a spare and true interchangeability.

OEMs today often have molding locations around the world to support their markets, which means that they may be sending the mold they're buying to another country for production. They want to know that anywhere in the world a cavity goes down, they can change out the down cavities with a new core/cavity, the parts will be identical, and production will continue.

Not all molds require such interchangeability. Assess your particular circumstances and work with your moldmaker to determine if you will benefit from having several sets of interchangeable cores/cavities, or what level of spares are required.[4]

TEXTURING

For cosmetic or aesthetic value, or to create a specific look for a finished product, components sometimes require a texture or decorative finish. To get this look, texture is engraved onto the surface of the mold cavity.

For example, a cosmetic case that holds face powder might have a shiny, smooth surface that indicates a tool with a high polish or "diamond" finish. A part in the interior of an automobile might have the look of leather using a texture on the mold surface.

There are dozens of textures to choose from, and the best way to select a surface finish is obtain test or sample plaques. These are standard finishes as defined by the plastics industry (see Bibliography). Some finishes the moldmaker can provide; however, others must go to an engraving or texturing vendor.

Texturing is generally quoted as a separate cost.

SOFT TOOLING

Soft tooling usually refers to molds made from aluminum or P–20 steel, and it can be less expensive than hardened tooling because aluminum or P–20 can be machined at much faster rates of speed. This reduces the number of hours required to construct a mold, and time is money. Tooling engineers will often budget a soft tool as a bridge, or intermediate, tool to:

- Obtain parts to use for evaluating the design, fit, or function of a new product.
- Test the properties of a product.
- Run preproduction parts while the multicavity production mold is being built.

Richard Caufman, a principal of RC Marketing Inc. in Warren, PA, offered some good advice about choosing a material for soft tooling in an article in *Injection Molding Magazine*.[5] He pointed out that there are many materials from which to choose, ranging from computer-generated resin models to specialty alloys. Here are some factors he suggests you consider when evaluating soft tooling.

Program life. Consider the life expectancy of the tool, the number of parts per order, the number of orders per year, and the number of years in the program.

How much aging will your tool experience because of startup stress? A mold that is run infrequently is more thermally stressed than one that runs continuously. Infrequently used tools, especially those molding high-temperature engineering resins, often make fewer parts than expected due to thermal stress.

Resin properties. Is the resin abrasive because it contains glass or metal fillers, is corrosive (like PVC), is flash prone (at .0001 inch), or is in need of a hot mold (more than 250F)? What kind of fill velocity and venting are required? What type of surface finish is needed for release and ejection?

Moldmakers know each increase in tool performance requires an increase in tool strength and precision. This necessitates a harder, or sometimes tougher, tool material, but not every moldmaker has experience with the material you will be using.

Part design. What kind and number of shutoffs are required? Each step away from a simple shutoff should be accompanied by a change in hardness or surface

Specifying a Mold 39

> **Moldmakers and Design**
>
> If your company has a design engineering staff, a moldmaker is often an excellent resource as an extension of that staff. Moldmakers can evaluate your part design, point out any problem areas, show you why the part might not be manufacturable in its current configuration, and help redesign the part for optimum manufacturability.
>
> For smaller companies, or for individuals who do not have access to in-house part designers, the moldmaker's design staff can offer valuable assistance in many aspects of design. Those that do not have designers on staff can usually provide names of firms or individuals who specialize in plastic part design and who can provide complete part drawings.

coatings. Also, ejection systems, side actions, and objects loaded into the tool (as for insert molding) influence the selection of tool material.

Productivity. What type of throughput is needed? Does adding 1 second to the cycle time cause accounting to go crazy? Is preventive maintenance a way of life or the slogan of the day? Given the life expectancy of the program, does it cost more to trim parts or build a better tool?

Maintenance costs. Who will pay for maintenance? Do you pay for it on an as-needed basis, or do you have a warranty agreement with the moldmaker? Since soft tooling requires more care than hardened-steel tooling, it can be that a soft tool will cost you more in the long run.

Aluminum molds are used as prototype molds in certain cases. However, some moldmakers prefer P–20 for what they term "pull ahead" or "bridge" tooling. Even though aluminum can be cut faster, cycle times are sometimes longer. A P–20 mold can be used to make production parts after approval, while the hardened-steel tool is being built, and then used as a backup tool.

For the sole purpose of shaving time off building the mold, high-speed machining technology has made the question of whether to use aluminum or P–20 almost moot. Even hardened steels can be cut at rates that were unheard of 10 years ago. Therefore, other considerations should play a role in your decision to go with a soft tool.

HOW MUCH WILL THIS MOLD COST?

How much does a car cost? If you're buying a Rolls Royce you'll pay more than if you're buying a Mercedes, and a whole lot more than if you're buying a Chevrolet. But a Chevrolet might be just what you need.

When specifying a mold, carefully consider your budget for the program and what percentage of this budget the mold represents. Budget constraints can also dictate the type of mold you buy.

For a large OEM, molds can be just one small component of the total costs-to-manufacture. For an individual buying a mold to produce a single product, the mold might be the most expensive component of the project. What return on your tooling investment are you looking for, and how long will it take you to recapture the dollars spent? You should understand what your limitations are in selecting a mold, and if the mold you need is going to fall within your budget.

Specifying a Mold 41

For example, you might go into the project thinking that you can get a less expensive aluminum mold. However, the material required to mold the part might be a glass-filled nylon, and you need a million parts annually. An aluminum mold probably isn't your best choice in this case. Let your moldmaker help determine if your needs fit your budget, and what can be done to help you meet your budget.

On the other hand, sometimes a more expensive mold is chosen when a less expensive one would work just as well. For example, a buyer might decide that it would less expensive to have a family mold—to build one mold with all the components together. In reality, a more cost-effective approach would be to have several molds with interchangeable inserts (a multiunit die) to mold the various components.

Remember, you have options. Sometimes it might be better to go with a more expensive mold up front (a hot runner mold, for example) to reduce manufacturing costs on the production end where you'll save in reduced material consumption, reduced scrap, and increased cycle times. You will then see a payback on the mold in a shorter time frame than if you had bought a less expensive mold that resulted in higher production costs.

A mold with a hot manifold system is more expensive on the front end than a standard runner mold. However, if your parts are conducive to this type of mold, you will probably save more money in processing by eliminating the runner system, which means less scrap and, sometimes, the need for an operator to clip the runners.

THE FOUNDATION OF YOUR PROJECT

You'll want to explore your options to determine which type of mold is best for your project. Involve a moldmaker from the beginning, and get his or her input. Build your budget around a mold that will provide optimum manufacturing capability in the most cost-effective manner, and be sure you're getting a mold that will meet all of your expectations for the life of the program.

The mold is the foundation of your entire project, and its success is critical to the success of the project.

Notes

[1] For more detailed definitions, see *What is a Mold?* by Tech Mold Inc., Tempe, AZ, p. 5–5.

[2] See "Monster Stack Mold Heads to Brazil," *Injection Molding Magazine* (Oct. 1998), p. 116.

[3] See "The Troubleshooter: Balancing Family Molds," *Injection Molding Magazine* (Jan. 1999), p. 94.

[4] See "Q+A: True Interchangeability of Tool Components," *Injection Molding Magazine* (June 1997), p. 68.

[5] See "The Selection, Care, and Feeding of Soft Tools for a Long Life," *Injection Molding Magazine* (April 1999), p. 83.

CHAPTER 3
PROTOTYPE PARTS AND MOLDS

Unanswered questions are everywhere when it comes to a new product. How does it look? Does it function the way it's supposed to? Will consumers like the way it looks? Will the design make for a good marketing tool? Such questions are one reason why people like to purchase prototype parts or molds before they commit to the production mold.

One item that builds cost into a mold is the "design as you go" concept (not to be confused with concurrent engineering), in which the designer of the component continually makes changes to the design even as the moldmaker is building the mold. This not only causes headaches for the moldmaker, but also adds expense to the project.

Marketing also plays an important role. Some firms will not go ahead with a new product until their marketing people have had a chance to take actual samples to their customers, get input on how well the product will be accepted, and decide how well form, fit, and function all work to achieve the purpose of the product. This process often requires only a few parts—maybe a dozen or a hundred or so—which means a full-blown production mold isn't really needed. This is where prototype parts and/or prototype molds come into play.

STEREOLITHOGRAPHY

Stereolithography (SLA) is a process in which a laser beam, following a pattern, solidifies liquid resin into a specific shape to form a basic prototype part.

Although stereolithography has filled a niche with respect to making prototypes in just a few days, be aware of a few drawbacks. For one, the part isn't made of the material in which the final component will be molded. While SLA provides a "touchy-feely" part that can be tested for form and design acceptability, the actual tests a product must undergo can't be accomplished because the properties of the SLA material and the final resin are quite different.

In recent years, new materials have been developed to give engineers more options. 3D Systems (Valencia, CA) originally produced three SLA systems that ran three general-purpose resins. By 1998, the company had expanded that to include 11 resins targeted for prototypes, tooling, and other uses. One new material, SL5220, gives moldmakers the high-heat resistance they need, along with dimensional accuracies within .001 to .003 inch.

HIGH-SPEED MACHINING

High-speed machining technologies, using CAD/CAM, have also created new opportunities for making prototype parts and molds in a matter of hours.

Why a Prototype Mold?

A prototype mold is used to test out design, run sample parts, check for fit and function, and provide samples for any product testing that is required prior to introduction into the market. With the new solids modeling software programs, new materials, and the use of CAD/CAM systems, prototype tooling can sometimes be built in a matter of days, instead of the weeks an aluminum tool often requires. In addition, prototype molds are generally less expensive and have fewer cavities.

When a customer of Arizona Precision Mold Inc., Mesa, AZ, needed a prototype part for a new style of portable telephone charger, APM first investigated using stereolithography. The prototype house quoted "three to four days and a fairly substantial cost."

APM came up with an alternative solution. Using its Pro/E software capabilities and a 3-D solid-model mockup sent as a CAD file via electronic data transfer, APM produced three sample parts machined from blocks of acetal overnight. The machined parts came within .001 inch of the customer's initial models, and allowed the customer to gain approval on size and shape.[1]

SLS VS. ALUMINUM

Another type of tooling that falls into the nonconventional category is produced by a process called selected laser sintering, or SLS.[2]

Paramount Industries, a turnkey product development firm in Langhorne, PA, embarked on a project for a customer in which it was able to perform a comparison of SLS vs. aluminum. It built duplicate tools of SLS using RapidSteel (a material developed by DTM Corp. of Austin, TX) and aluminum. Paramount eventually did the five-part project using RapidSteel 2.0 and copper polyamide (another DTM material) materials, as well as the traditional CNC machining and aluminum, doing all three in parallel. RapidSteel, which grew out of DTM's earlier material, RapidTool, proved superior in that it needed less finishing time than the earlier material. Additionally, RapidSteel 2.0 also reduced total lead time from 5.1 weeks (aluminum) to 4.5 weeks.

Within five weeks, the customer had injection molded ABS and elastomer parts produced from both machined aluminum tooling and SLS-grown RapidSteel 2.0 tooling. The parts looked nearly identical. Tolerances on SLS-grown tools are not as accurate as on aluminum tools for all dimensions. After checking the prototype parts for fit and function, the customer made only a minor design adjustment to one of the parts. Some time savings were realized; however, Paramount estimated that turnaround time for the project would have been reduced by one week had the SLS process been used independently— and that could have been improved with overtime. The time frame included build, sample, secondary work, second sample, and first-article inspection reports.

An article in the September 1999 issue of *Injection Molding Magazine*[3] details more successes of the SLS process, which can allow moldmakers to produce a mold in days rather than weeks, and give the customer parts in the specified material. In some cases, as many as 50,000 high-quality parts have been produced in filled engineering resins in rapid tools. Runs from 10,000 to 20,000 parts in materials like glass-filled nylon and polycarbonate are common.

Rapid Solutions, a company in Nashotah, WI that is pioneering the use of SLS rapid prototyping technology, delivered $15,000 worth of tooling for molding cosmetics containers in just nine days. Although tolerancing with SLS is still not as predictable as it is with machined molds, people in the rapid prototyping/rapid tooling industry are confident that new developments in design software, sintering technology, and materials of construction will improve accuracy while opening the door to new tooling opportunities.

ALTERNATIVES

A good bridge between prototype tooling and hard tooling has been developed by Tupperware in Orlando, FL and Vintage Industries in Longwood, FL.

Tupperware engineers needed a way to test the non-spill mechanism of a new children's spill-proof cup, but didn't want to go the expense of a hard tool. Their answer for their Sipper Seal project was a thin-shell nickel electroformed over an SLA pattern, and then back filled with a low-melt alloy to produce core and cavity inserts.

Their success led Vintage to create a division called Nickel Composite Tooling, which produces nickel mold inserts for either production or prototype molds, depending on volume levels. One big advantage is that the nickel-shell tool can maintain dimensional stability in production runs up to 500,000 parts.[4]

Sipco Inc., a moldmaking company in Meadville, PA, has developed a way to shave weeks off standard lead times. Kevin Maziarz, sales manager for the 40-year-old company, says, "Rather than building dedicated prototype or prehardened tools, we encourage our customers to think production from day one."

The company builds a multicavity base, and then "pulls ahead one cavity steel-safe." This means the company will push through design and build on just one of the cavities, and make it so that changes can be made to the component and the cavity as engineering continues to perfect the design.

"Our customers are getting samples from a production tool in the same amount of time it would take to build a prototype," notes Maziarz.[5]

Stiff competition in both price and delivery lead times are forcing moldmakers to become more creative in the ways they help their customers achieve time-to-market goals. Ask your moldmakers to quote programs on how to help you shave time off the project. Make this a consideration when making your final selection of a moldmaker to do your job.

HARDER AND FASTER

Finding a way to make relatively hard tooling more quickly than cutting aluminum was the driving force behind a research partnership between JCI-Prince, an automotive components supplier in Holland, MI, and

Ciba Specialty Chemicals in East Lansing, MI. The result was the development of an aluminum-filled thermoset composite board that can be machined in hours rather than days or weeks, that eliminates the need for stereolithography pattern masters, and that requires no secondary finishing.

Called Cibatool-Express material, the board is commercially available and, unlike aluminum, the composite inserts manufactured by the companies need no secondary finishing or surface treatment. Using high-speed machining equipment, the composite-board tools are produced in "15 to 20 percent of the time required for aluminum, yet part quality and dimensional accuracy were identical."[6]

CHOICE OF PROTOTYPE

Whether you choose to make prototype parts via stereolithography, a prototype mold in soft steel, or using the Rapid Tool process depends on many factors of the project. What is your time-to-market window? What are your budget considerations? Do you need a soft-steel tool to use as a "pull ahead" production tool, or are you confident enough to move from prototype parts to hard-steel production tooling?

Options exist to help you get to where you need to be. Prototype parts and molds can help you answer many of the engineering questions along the way, resulting in an optimum product at the end of the line. Talk to your moldmaker about your requirements and evaluate those options.

Notes

[1] See "Stretching the Limits of Simulation for Molds and Prototype Parts," *Injection Molding Magazine* (Sept. 1998), p. 85.

[2] See "Comparing SLS and Aluminum Tools," *Injection Molding Magazine* (Oct. 1999), p. 77.

[3] See "Prototyping Specialists Span the Range," *Injection Molding Magazine* (Sept. 1999), p. 46.

[4] See "Thin-Shell Nickel Tooling Seals Up Time Savings" (Jun. 1998), *Injection Molding Magazine* (June 1998), p. 71.

[5] See "Equipped for Competitive Speed," *Injection Molding Magazine* (July 1999), p. 65.

[6] See "Rapid Tools for Prototypes and Beyond," *Injection Molding Magazine* (Aug. 1998), p. 82.

CHAPTER 4
REQUESTING A QUOTE

When you have a good idea of the type of mold to specify for your part, you are ready to prepare a request for quote (RFQ). Keep in mind that most mold shops get dozens of RFQs every week—sometimes those involved in the quoting process say they are drowning in RFQs!

Quoting is a time-consuming process considered by most moldmakers to be a nonvalue-added cost of sales. That is, unlike other functions in a mold shop such as designing or machining—functions that add value to the product—quoting gives zero return on investment. Unless, of course, the moldmaker gets the order for the job.

The cost of quoting is often considered a cost of sales at best, and for many moldmakers, that cost falls into the black hole of accounting.

HOW MANY IS TOO MANY?

Soliciting too many bids is a waste of everyone's time and shows a lack of knowledge on the part of buyers about mold suppliers, according to William J. Tobin, WJT Assoc., Boulder, CO. In his book *Injection Mold Tooling Standards,* Tobin suggests that if you have doubt about which mold shop is capable of making the mold you need, enlist the assistance of a professional consultant to help you make a choice.

It does take research to locate those moldmakers who can provide the technology and expertise in designing and building the specific type of mold you need. (Advice on finding appropriate moldmakers is in Chapter 5.) However, several resources are available, including industry organizations such as the American Mold Builders Assn.(AMBA), headquartered in Roselle, IL.

Visit a shop before sending it an RFQ for a mold that might be out of its realm. Not only will you save yourself time, you cut short any temptation the moldmaker might have to try and build a mold he's not capable of or doesn't have the time to build.

If you have done your homework and located three mold shops that, from your research, appear to have fairly equal capabilities and expertise, then three RFQs should be enough for a realistic picture of what you can expect. Responses to a dozen RFQs could possibly give you a distorted picture of your mold program's true costs.

THE QUOTING PROCESS

Considering the volume of requests that come through the door, many shops often like to prequalify their RFQs. Therefore, you could get a call asking questions about your RFQ. When is the project being released, at what stage of development is the project in, is this a ballpark quote for marketing for some cost estimating or for a feasibility study for the component, or is this the real thing? Is there an existing mold? If so, what do you like or dislike about the existing mold?

Prequalification of RFQs is done because most of the quotes moldmakers supply are not just ballpark figures, thrown out after a cursory review of the print. Prequalifying your RFQ helps the moldmaker determine whether or not this is the right type of job for his shop, and if it fits into his work schedule. Additionally, he wants to make sure there is a likelihood that the project will be awarded in a reasonable amount of time, and that the hours he puts into his RFQ are not a waste of time.

Quoting usually entails engineering evaluations in which the part prints are gone over carefully to ensure nothing is missed that might cause costly or time-consuming complications down the road. Missing a detail can be disastrous for both the moldmaker and the customer, which means that varying levels of quotes may be called for.

For example, if you are uncertain about the feasibility of your part design and want an engineering evaluation of its moldability as drawn, then you should ask specifically for an engineering evaluation of the design as a precursor to the quote. An engineering evaluation takes longer than a basic ballpark quotation for an as-is design.

However, if you need a ballpark quote for budgetary purposes or to explore the feasibility of making the component, by all means request a generalized estimated-cost quote. Many moldmakers claim that even the best, most accurate quote is a guesstimate that can prove high or low, depending on other factors that can arise downstream. That's why it's good to specify whether you need a preliminary ballpark quote or a complete engineering evaluation.

STANDARD PRACTICES

It is standard practice in the moldmaking industry to supply a complimentary engineering evaluation if the mold vendor is preselected (usually in a partnership arrangement), or if the moldmaker is already a preferred supplier for the customer requesting the evaluation. That way, a moldmaker is guaranteed that the time spent on performing the evaluation will be repaid through the mold build.

If an engineering evaluation is requested as part of the quote process prior to the selection of the moldmaker, then a charge may be applied to cover costs. All too often, a moldmaker's early input on design changes and optimum tooling design helps only his lower-priced competitors, which leads us to a discussion of intellectual property.

RESPECTING INTELLECTUAL PROPERTY

An engineering evaluation of a part design and recommendations on optimum mold design are the product of a moldmaker's years of expertise and the creative talents acquired from that experience. As he reviews a part print for quoting purposes, he may see specific areas that he believes can be enhanced if certain changes are made. Many times he will include notations of "good idea" or make design suggestions that can help the customer. The engineering evaluation therefore contains a moldmaker's intellectual property.

It is unethical for a purchasing agent or engineer from an OEM to take the recommendations and engineering changes supplied by one moldmaker and give them to another moldmaker whose bid might be lower because he didn't recognize problem areas or situations that would require a more complex mold design or certain changes.

Not only does it hurt the moldmaker who took the time to provide the engineering evaluation and make the recommendations, it's also a disservice to the moldmaker who gets the job, because now he's locked into a price and faced with a whole set of changes to the part design and mold design. This can also put the quality of the your product at risk.

The goal should always be to obtain the best possible mold from the most qualified supplier at a fair price. A part design for a component or a product belongs to the owner of the component or product. The mold design belongs to the moldmaker.

Mark S. Mahoney, an attorney in Pacifica, CA, works with molders and moldmakers as well as plastics industry trade groups on legal issues affecting the industry. He suggests that moldmakers have a Confidential Disclosure Agreement, signed by both the moldmaker and the customer at the time of the engineering evaluation or quote. This could be similar to an agreement Mahoney developed (see Table 4.1) in response to questions about this issue from the American Mold Builders Assn.

Table 4.1 Confidential and Proprietary Information Nondisclosure Agreement (*Source:* Mark S. Mahoney)

> This agreement (the "Agreement") is made between The X.Y. Smith Company ("Moldmaker") and _____ ("Customer" or "you") and entered into this ____ day of _____, 20__. In consideration of the mutual promises and covenants contained in this Agreement, Moldmaker's disclosure of confidential information to Customer, the parties hereto agree as follows:
>
> The purpose of this Agreement is to protect the proprietary property rights of Moldmaker to its designs and research that are developed by Moldmaker for Customer. Moldmaker's confidential and proprietary rights include the following: product design and redesign for parts, molds and related tooling, slide configuration, cooling configuration, injection and ejection design, . . .
>
> Moldmaker shall take possession of Customer's documents and design for the purpose of professional review and analysis. Moldmaker shall supplement Customer's design with Moldmaker's own proprietary technical information, expertise, research, and development, which are mutually acknowledged to be Trade Secrets.
>
> Customer shall not disclose the information supplied by Moldmaker without prior written approval for a period of five years from the date of execution of this Agreement. In the event Customer elects to obtain moldmaking services from another, a competitor of Moldmaker, Customer agrees to pay Moldmaker a fee of $xxx per hour for the research and development costs. This fee shall not waive or release any duty of Customer to maintain the confidentiality required by this agreement.
>
> Customer shall not use such information other than for mutual review and analysis for the manufacture of molds and related tooling by Moldmaker. Customer agrees to return any documents disclosed by Moldmaker immediately upon request.
>
> Any action to enforce this Agreement, the terms of this Agreement, or any action related to the relationship between the parties to this Agreement shall be conducted in Moldmaker's Own Home County and according to its State laws.
>
> ALL PROPRIETARY INFORMATION SUPPLIED HEREUNTO IS SUPPLIED "AS IS" AND MOLDMAKER SHALL NOT HAVE ANY LIABILITY FOR DEFECTS IN PROPRIETARY INFORMATION SUPPLIED HEREUNTO. MOLDMAKER SHALL NOT BE LIABLE FOR ANY DIRECT, INDIRECT, SPECIAL, INCIDENTAL, OR CONSEQUENTIAL DAMAGES OR LOSS OF USE, REVENUE, OR PROFIT ARISING OUT OF THE USE OF PROPRIETARY INFORMATION FOR THE PURPOSES OF THIS AGREEMENT. THERE IS NO WARRANTY OF FITNESS FOR A PARTICULAR PURPOSE.
>
> Date:_____   Moldmaker:_____
>
> Date:_____   Customer:_____

HOW TO GET THE CORRECT QUOTE

A term frequently tossed about in mold buying is "Class A mold," but no industry standard exists for a "Class A" mold. The term is usually misused by potential buyers when they want a high-quality mold; however, it doesn't tell the moldmaker anything about requirements and leaves too many questions unanswered.

The Society of the Plastics Industry (SPI), headquartered in Washington, DC, provides industry-recognized classifications of molds (see Bibliography). These classifications are based on the estimated number of parts that specific types of molds, made from various types of metals, can be expected to run over the life of the mold.

If you have already determined that the part print or database you are sending is viable, then you are ready to request a quotation based on the print's design and other data to be supplied to the moldmaker. But leaving it up to the moldmaker to make any type of mold he feels is OK for your part opens the door to problems.

For example, without specific guidelines as to exactly what type of mold you need, each moldmaker who is quoting the job will quote according to his expertise, experience, and capabilities. You may send out an RFQ package of four parts that are all to be made of the same material and similar in size. Moldmaker A might quote you a family mold. Moldmaker B, who doesn't like family molds, might quote you a multiunit mold. And Moldmaker C might quote you four completely separate tools.

Thus, for three RFQs you could get back three quotes for different types of molds, configured in different ways, and with pricing all over the map. Additionally, if the moldmaker builds a mold the way he feels is suitable to mold your part, and problems arise later, who's to blame?

One way to avoid these problems is to supply a detailed request for quote filled out according to the Mold Quote Form (Table 4.2) that is printed in the Society of the Plastics Industry's publication, *Customs and Practices of the Moldmaking Industry*. This form will help you be very specific in the type and quality of mold you want.

Thus, the moldmaker will know exactly what you want. You will have eliminated guesswork, and fewer problems will await you down the road.

BEING SPECIFIC

Being specific on an RFQ helps you obtain apples-to-apples quotations. To accomplish this in a way that gets you the best quote possible means you must provide as much precise information as possible.

During his career as a tooling purchasing engineer for a variety of industries, including automotive, consultant Bill Tobin ordered hundreds of molds. Today, Tobin believes there is a formula for the perfect RFQ. His first rule of thumb: Write a request for quote that specifies exactly what you want.

"Moldbuilders are not mind readers," he says, "so the RFQ must be as clear as possible."[1]

It's not what's in the RFQ that causes problems, he continues, but what's not. Tobin points out, correctly, that mold buyers are not buying molds, they are buying capacity, and most moldmakers will not guarantee or even quote the specific cycle time they expect their molds to run. This is because they believe they have no control over how the mold will be run, in what press it will run, or how the setup person and operator will establish molding parameters.

"Somewhere in the RFQ must be a statement that the tooling is capable of producing the expected quantity of parts per year with reasonable maintenance and reasonably predictable downtime for machine maintenance," he notes.

Here is some of the information that should be included in an RFQ for a mold. (Along with the mold quotation form, SPI also offers guidelines for tooling classifications and the number of parts a specific classification can be expected to produce[2].)

a. Part name, number, any pertinent engineering revision numbers/letters, date print was released for bid.
b. The expected lifetime production numbers of the part and cavitation required.
c. The type of plastic material to be used and the expected number of parts to be molded annually and/or the number of parts required weekly/monthly.[3]
d. Texture on any mold surface (type and exact area to be textured), platings required.
e. Classification of mold (the various mold classifications are important, as not every mold will need to be a "Class A" mold, and, as has been pointed out, there is no industry-recognized definition of "Class A").

Table 4.2 Mold Quote Form (*Source:* Society of the Plastics Industry)

Customer Name _____ Date _____

RFQ# _____ Quote # _____

Name 1. _____ Part No. _____ Rev. No. _____ No. Cav. _____
Of   2. _____
Part/s 3. _____

Total No. of Cavities _____

Type of Mold ____ Thermoplastic injection ____ Compression ____ Transfer ____ Other

Design by   **Type of Design**   **Approximate Mold Size**
____ Moldmaker   ____ Detailed Design   Press _____
____ Customer   ____ Layout Only   Make/Model _____

Mold Construction   **Mold Base Steel**   **Material**   **Hardness**
____ Standard   #1____ #2____ #3____   Cavities   Cores   Cavities   Cores
____ 3 Plate   S.S.____ Other____   ____ Tool Steel   ____ Hardened   ____ 50-58 R
____ Stripper   ____ Beryl. Copper   ____ Pre-Hard   ____ 28-32 R
____ Insulated Runner   Inserted   ____ Hobbed   ____ Other   ____
____ Reverse Ejection   Primary Core Yes____ No____ ____ Other
____ Runnerless System   Primary Cavity Yes____ No____

Table 4.2 Mold Quote Form (Continued) (Source: Society of the Plastics Industry)

Type of Gate

- Edge
- Center Sprue
- Subgate
- Pin Point
- Hot Bushing
- Post Gate
- Other

Ejection

Cavities		Cores
___	EJ Pins	___
___	EJ Blade	___
___	Sleeve	___
___	Stripper	___
___	Air	___
___	EJ Bars	___
___	Unscrewing (Auto)	___
___	Removable Inserts (Hand)	___
___	Other	___

Special Features

- Guide Ejectors
- Spring Loaded EJ Bar
- Accelerated EJ
- Positive EJ Return
- Cylinders on EJ Bar
- Parting Line Locks
- Double Ejection
- Other
- Threaded EJ Bar

Side Action

Cavities		Cores
___	Mechanical Slide	___
___	Hydraulic Cyl	___
___	Air Cyl	___
___	Cam	___
___	EJ Activated	___
___	Spring Activated	___
___	Angled Lifters	___
___	Collap. Core	___

Cooling/Heating

Cavities	Core
___ Blocks	
___ Mold Base	
___ Other	
(Specify)	

Finish

Cavities		Cores
___	SPI#	___
___	Mold Base	___
___	Texture	___
___	Other	___

- Heaters Supplied By _____
- Heater Connectors By _____
- Outside Surface Model By _____
- Duplicating Casts By _____
- Mold Tryouts By _____
- Matl. Supplied By _____
- Moldflow Analysis By _____
- Cooling Analysis By _____
- Controller Supplied By _____
- Molded Part Inspection By _____
- CAD/CAM Data Supplied By _____

Engraving

_____ Yes _____ No

Micro Switches

Moldmaker _____ Molder _____
_____ Supplied By
_____ Mounted By

Moldmaker _____ Molder _____

SIGNED _____

f. Whether the mold will be a hot runner mold or a cold runner mold. If a hot runner mold is needed, who supplies the hot manifold controller.
g. Spares required.
h. In what size press the mold will be used.
i. Whether steel-safe recut is to be used. Whether steel inspection is required.
j. What certifications are required.
k. What level of sampling of the mold is required—first article only or full process qualification and pre-production run.

Allowing the moldmaker to guess what type of mold is best for your project can also lead to problems down the road if a component you wanted in the mold isn't originally quoted. Then the situation becomes a finger-pointing contest because you have to stick to the quoted price for budgetary purposes and you think the moldmaker should add the component(s) at his cost. Almost certainly, the moldmaker will not want to pay for the additional time and materials to correct a problem created because your RFQ was not specific enough.

On the other hand, requesting quotes for a wide variety of cavitations and configurations is problematic for moldmakers, and a situation they openly complain about when discussing RFQs. For example, a request might be made for a part to be quoted in a two-cavity, four-cavity, and eight-cavity mold with conventional runner and hot runner systems, and for quantities ranging from 1000 to 1 million parts annually.

What It Takes

Providing a valid quote requires the involvement of several people, usually including the shop's mold designer, a salesperson, and perhaps the manager. If a custom molder is to do the molding, someone representing the molder will also be involved so that processing parameters can be included.

Some shop owners have been known to take RFQs home and stay up late poring over blueprints to get the RFQ out on time. It takes an average of 2 to 4 hours per print/part to examine blueprints, note exceptions, identify problem areas, ensure that all the details have been covered, and then come up with a viable cost estimate.

Or, when several different parts are being quoted, the request might be to quote them in multiunit dies with interchangeable inserts, in a family mold, or in individual molds. This means that the RFQ takes longer to complete, and it also tells moldmakers that you don't exactly know what you want in a mold—something that tends to make them nervous.

If in fact you don't know what type of mold you will require, instead of sending an incomplete RFQ or one requesting a variety of configurations, work with a moldmaker and pay for an engineering analysis of your component and its molding requirements to nail down those details before sending an RFQ.

The more exacting and specific you can be about what you really need quoted, the less time it will take to provide the quote and the more likely you are to get an accurate estimate.

EXCEPTIONS

Don't expect or assume that a moldmaker will note design errors in the quote stage. Moldmakers generally spot small errors or note potential problem areas that could interfere with optimum processing, but unless the mold designer has been involved in the development of the part from conception, it is unlikely he will do any redesign of the part.

In the past, moldmakers were reluctant to make exceptions on a mold due to an error in part design. They felt that it put too much responsibility on the moldmaker for the part's design, which could lead to liability issues down the road. That is changing, however, as mold shops add new product development assistance to their offerings. These designers work with the product development team to ensure that the components are designed correctly from the outset—a real time and money saver down the road.

However, shops will generally note any exceptions and either quote a mold with the exceptions included in the price or provide an "as is" quotation with exceptions noted.

WHY THE BID IS THE BID IT IS

When you receive a quote, it can be confusing unless you understand why the moldmaker has quoted the job as he has. Every moldmaker to whom you send an RFQ will have a slightly different idea of how your specific mold should be built, even if they have all been given the same specifications.

In addition to industry standards that ensure molds are built to specification, moldmakers bring to the table their own creativity and years of experience. Often that is what's going on when a purchasing agent gets back three quotes from three different moldmakers, and the prices range from very low to very high to somewhere in the middle, in spite of the fact that all three moldmakers were working from the same information.

Some purchasing agents automatically throw out the lowest bid, believing that it can't possibly be a good mold if it's that cheap. Generally, that is true. (The old adage, "If it seems too good to be true, it probably is," should be kept in mind.)

On the other hand, a low bid might be a bargain. It is well known in the industry that when mold shops are busy their quotes are higher. Moldmakers rarely refuse to quote a job—even if they are extremely busy—for fear that the customer won't offer them another opportunity to bid. So, they simply quote a higher price and/or a long delivery time as a way of discouraging the buyer from giving them the job, while at the same time preserving their standing for future opportunities. (Some purchasing agents would rather have a busy shop submit a "no quote" than to quote very high. If you feel this way, make your preference known to the moldmaker.)

Never assume that just because a price is low in comparison to other bids that it is a bad quote. However, it is important that you contact all the moldmakers who submitted bids and discuss their prices (see "Offer To Negotiate" later in this chapter). Get a realistic picture of why they quoted a particular mold as they did, and determine the cause of the difference in price, particularly if that difference is significant.

It also helps the moldmaker to know who he is quoting against. Understanding if he is on a level playing field with respect to his competition influences the way he will quote a job. This is another reason that it is critical to prequalify the moldmakers you are considering even before you send an RFQ. It's difficult to evaluate a quote when the shops providing the numbers are extremely different in size, technology, expertise, and capabilities. In Chapter 5, we discuss in more detail how to select the right moldmaker for the job.

USING 3-D CAD TO ESTIMATE MOLD COST

3-D solids modeling influences the way in which some cost estimating of molds is done. Some moldmakers routinely receive 3-D CAD models of the plastic components on which to base their quotes instead of the typical 2-D drawings.

However, Delta Tech Mold (Arlington Heights, IL) found that this was actually adding time to its estimating process.[4] Customers were sending 3-D models using different CAD software from Delta Tech's, and a lot of time was being spent translating the files to produce the needed dimensioned drawings.

The company solved the dilemma by purchasing a solid model viewing package that reads IGES, STL, DXF, and VRML files, and then performs real-time shading and rotation with a standard PC. Not only was the estimating time reduced, but the system provided estimators and engineers with complete 3-D dimensioning of edges, faces, arcs, vertices, and completed objects, resulting in more accurate estimates. The company installed Internet capabilities so that files could be sent electronically.

Better viewing of the part allows better understanding of the geometry involved, removing uncertainties from the quoting process and resulting in more accurate pricing, notes one of Delta Tech Mold's customers.

HOW A QUOTE IS FIGURED

How does a mold shop arrive at a price? To begin with, the part print is examined carefully, usually by several people. These people are not only looking for any problem areas that might create complications in the mold build, but they are also exploring options. What are the possible scenarios for a mold for this particular part based on the OEM's requirements? What is the best mold for this part to allow for optimum manufacturability and highest productivity and efficiency?

The creativity of the individuals involved in the quoting process plays a significant role in arriving at the numbers. They mesh their answers to these questions with the specific requirements outlined on the RFQ, such as the type of mold, the components it requires, the raw materials needed, and the estimated hours that will be involved in building the mold. Then they figure out a price.

The price that is quoted also depends on whether you are dealing directly with the moldmaker or with a custom molder who has in-house moldmaking capabilities but whose own mold shop is too busy and so decides to subcontract the mold outside. (Custom molders that do not offer in-house moldmaking services often act as an intermediary and purchase the mold for the customer. These molders generally add a percentage of the total cost of the project to the price of the mold to reflect their management time in overseeing the project.)

If you, as a mold buyer, are a custom molder working on behalf of an OEM customer, keep in mind that everyone is price-sensitive these days. It's a good idea to work with your moldmaker to determine appropriate fees for your mold management time, so that the price is in line with your customer's budget. It may be that your customer doesn't object to the additional cost incurred by having you manage the mold build with an outside moldmaker. Be sure to discuss this in advance.

The point is, from whom the quote is coming—a molder subcontracting the mold build or directly from a moldmaker—affects the bottom line price. Take this into consideration when evaluating quotes.

UNDERSTANDING THE QUOTE

Now that you have a quote in hand, what does it actually tell you? What it tells you can only be as thorough as the information you provided in your RFQ (Figure 4.1).

Usually moldmakers break out the individual components of the mold build like this:

Job	Estimated time	Estimated cost
Clean up database	20 hours	$0000
Mold design	2 weeks	$0000
Mold build	8 weeks	$0000
Total	21 weeks	$00,000

Requesting a Quote

Figure 4.1 Sample Mold Quote (*Source:* Tech Mold)

TECH MOLD, INC.
1735 West 10th Street
Tempe, Arizona 8521-5295
Phone: (555) 555-5555
Fax: (555) 555-5555

XYZ CORPORATION
123 PROFIT ROAD
ANYTOWN, USA

RFQ. NO: __XYZ1__
F.O.B.: __Our Plant__

DATE: __00/00/00__
PRICE VALID: __30 Days__

DESCRIPTION	32 CAVITY WIDGET MOLD	
DESCRIPTION OF WORK		PRICE
Hot Manifold directly feeding part Hot Tip Gating Parting Line Straight Locks Chrome Plating Mechanical Slide Actions Guided Ejection Stripper Plate Ejection Engraving: Cavity I.D., specific verbiage and logo Cores Finish - A-2, Grade #6 Diamond Cavities Finish - A-2, Grade #6 Diamond		
Cavity stack-up made from standard hardened tool steels. Standard tool steel selection to consist of the following choices; 420 SS, A-2, A-6, A-10, S-7, H-13, M-2.		
Non-standard tool steel selections required by either the customer or design parameters may result in added charges that will be addressed during design.		
Cavity & Core Cooling - Circ. around Insert & Fountains in Cores		
Finished plan views and section views. Complete bill of materials. All cavity components detailed and tolerenced. Stock purchased components not drawn unless they are to be modified. Each component drawn is to be on an individual sheet. Each mold plate to be on an individual sheet.		
	Design - 5 weeks	XX,XXX
	Build - 15 weeks	XXX,XXX
	Mold Evaluation Sample	X,XXX
	25% Spares	XX,XXX
	Steel Certification	XX,XXX
	Sub Total -20 weeks	XXX,XXX
	Steel Safe Recut - 2 weeks	XX,XXX
	Sampling After Recut	X,XXX
	Total - 22 weeks	XXX,XXX

Other charges could include mold sampling and first-article inspection,[5] although usually these are provided by the molder you've chosen to mold your components. Some mold shops have in-house mold sampling capabilities; others will send the mold to a local molder for sampling so that any corrections can be done quickly. Or you may choose to have the mold shipped to your molding facility for sampling and first-article inspection.

The detailed information on the quote allows you to see the breakdown of the program in terms of how long each phase will actually be, which is of great assistance with your planning. If the moldmaker does not break out costs in this manner—and some do not—you can always ask him or her for a breakdown. Also on the quote should be the details of the mold following the guidelines of the RFQ. Exceptions will be noted as mentioned above.

After reviewing all the quotes you receive, you then determine which price and delivery time meet your budget and schedule. At this point, you may have to negotiate with your potential moldmaker for any factors you consider not acceptable.

Leveling the Workload

One of the most difficult aspects of operating a moldmaking shop, from a management perspective, is keeping the workload on an even keel. Like many job-shop environments, moldmaking shops are either buried with work or folks are standing around waiting for a buyer to release the purchase order for a big job.

Therefore, a hungry moldmaker tends to quote lower to help ensure getting the work. So some shops really might be offering you a bargain with a bid that is lower than its competitors. Get to know the shop and its capabilities, and understand why they quoted low in order to be sure that they didn't miss seeing something on your print that might cause a problem later.

OFFER TO NEGOTIATE

Generally, moldmakers are open to negotiating a bid. If a moldmaker isn't familiar with your company, or hasn't built a mold for you in the past, he'll probably cover the unknowns by being a bit high in price. Offering to work with him on the type of mold you need, as well as on terms and conditions, will help establish a comfort level with you and your company, making him more willing to negotiate.

Keep in mind, however, that moldmakers seldom put a lot of fat in their quotes. Most have been building molds for years and know what their costs are, and how much time it takes to build certain types of molds. Don't expect a moldmaker to reduce the quoted price by a large percentage. As moldmakers like to say, "If I was able to shave 20 percent off the price, I wouldn't have quoted that high in the first place!" Moldmakers attempt to quote competitively to get the work.

THE IMPORTANCE OF LEAD TIME

Lead time, or delivery date, is quoted based on a shop's capacity, how busy it is at the time of the quote, and the capabilities of its people. Like price, lead times will vary widely.

One factor affecting a promised delivery date is how long it will take for the mold buyer to give the go-ahead after receiving the quote. OEMs are often slow to release purchase orders—marketing hasn't finalized the design, money for the project hasn't been approved, or the project has been put on hold until more research and development is completed.

> **A Reasonable Amount of Time**
>
> The usual time to turn a quote around is a week to 10 working days. Most moldmakers try to accommodate their customers' requests with respect to meeting these deadlines; however, a quote can be done more quickly if you have a legitimate need for it sooner. Because some buyers wait until the last minute to send an RFQ, they think it's OK to impose a deadline of a 24-hour turnaround. That's unreasonable.
>
> If you are a regular customer, mold shops are much more inclined to provide quotes quickly when you are up against a deadline. However, given the sheer volume of RFQs, most shops try to keep rush quotes to a minimum. Some moldmakers claim that every RFQ they get is marked "urgent," which means they couldn't possibly provide 24-hour or 48-hour turnaround on all of them.

In the meantime, the mold shop's workload has picked up considerably, which in turn affects the original delivery date stated on the quote. Bill Kushmaul, president of Tech Mold in Tempe, AZ, notes that customers are often under the misconception that once a mold shop quotes a delivery time frame—six weeks, for example—it will then wait for the customer to release the purchase order, even if that doesn't happen for another month following return of the RFQ.

"We need to emphasize more that our business changes on a daily basis," Kushmaul says. "One week we're slow and looking at available machine time. The next we've got so many purchase orders coming in the door that our lead times begin to look impossible."

Some customers tend to look for shops that are empty, he explains, hoping to get a fast delivery but not realizing that an empty shop may or may not be an efficient shop. A shop that is maxed out isn't necessarily efficient either, and that's when the issue of costly overtime becomes critical for moldmakers—and mold buyers.

Most mold shops try to have some backlog at all times, because they can't afford to have high-paid moldmakers standing around waiting for a purchase order.

Large companies often release purchase orders when budgets are released, such as at the beginning of the calendar year. This results in moldmakers becoming extremely busy at that time. Other companies release purchase orders for molds at the beginning of a fiscal year, or when marketing and management have approved the program. Contracting for molds during off times can work to your advantage.

Before locking in your program on a delivery date based solely on a quote you received weeks, or even months, earlier, check to see if any changes have occurred in the moldmaker's workload.

Negotiating dates, as well as price, can go a long way toward preventing glitches.

THE RFQ SAYS IT ALL

An RFQ asks for a bid. A quote is the offer. Ultimately, your purchase order is the acceptance of that offer. Because the RFQ is the first step toward building a contractual relationship with your moldmaker, it should say everything needed to build this contract.

Look at your RFQ as the critical link between the mold you need and the mold the moldmaker will build. Knowing what to ask for, and then asking for it, is what your RFQ is all about.

Notes

[1] See "The Perfect RFQ," *Injection Molding Magazine* (July 1999), p. 18.

[2] Society of the Plastics Industry, *Customs and Practices of the Moldmaking Industry*.

[3] There is some question as to who should make material recommendations. Moldmakers need to know the type of plastic you will be running because plastic resin will shrink at different rates in molds of different cavitation and type. Resin producers have specifications on their products; however, the range of shrink they offer tends to be wide to cover themselves in the event of any liability issues. Moldmakers and molders are hesitant to specify the material for the same liability concerns. Most moldmakers have developed an expertise in determining the shrink rates for various molds, but require that the person purchasing the mold make the final decision based on input from the resin supplier, the molder, and the moldmaker.

[4] See "Estimating Mold Cost With a View," *Injection Molding Magazine*, (April 1998), p. 82.

[5] The first parts out of a new mold, or a mold that has had an engineering change or rework, usually undergo an initial inspection of all tolerances to ensure they are within specification.

CHAPTER 5
CHOOSING A MOLDMAKER

Now that you have your quotes gathered, the next step is to get your mold built. This means selecting the right moldmaking shop for your job.

If you've been involved with the plastics industry for any length of time, you already know that there are several ways you can go about getting molds built. You also know that, in most instances, you have dozens of mold shops to choose from, and that finding the correct shop to build the mold you require is not always an easy task.

No one really knows exactly how many moldmaking shops are operating in the United States. Some industry guesstimates place the number at 5000, more or less (not counting all the molding companies that have in-house moldmaking). However many, lots of moldmakers are out there, as you'll discover if you look through your local Yellow Pages or in the Thomas Register (either the book or the online version at www.thomasregister.com). You can also locate hundreds of moldmakers on the Internet by searching for "mold" or "injection molds."

Mold shops range in size from one or two people and a few pieces of equipment to 100 or more machinists, design engineers, and moldmakers. According to the American Mold Builders Assn., the average size shop of its members has about 17 employees and sales of about $3 million annually. In other words, the typical mold shop is still a small business, privately held, and, in many cases, with a single owner or two partner owners.

When you call a mold shop and ask to speak to someone about a mold, frequently you'll be put through to the owner. Generally, he was once a moldmaker or designer who, as business grew, was relegated to the front office for sales and to keep an eye on operations. Some larger mold shops have sales engineers or sales reps, but to keep new work coming in, many still depend on word-of-mouth or repeat business from long-term customers.

It's been said by people who work in the industry, and even by moldmakers themselves, that they are great at making molds but less savvy when it comes to business. However, in recent years moldmakers have become increasingly sophisticated about business issues as they learn to play on the same field with the global corporate entities.

THE MOLDER IN THE MIDDLE

Numerous molders and moldmakers have formed strategic partnerships and worked well together for years on projects for mutual customers. These successful relationships are those in which communication pitfalls have been overcome, and each knows what the other needs to do his part of the project successfully. However, in other situations, where a molder is the "middleman" between the mold buyer and the moldmaking shop, mistakes can easily be made when information is not passed along from one party to the other.

Even when a moldmaker deals directly with the OEM, the molder should be involved because processing factors should always be considered, such as the press that will be used, cycle times, and so forth. These are factored into the part price, which makes it critical for the molder to have input on the mold design.

Custom molder Dave Brentz, vice president of sales and marketing for Plastech Corp. in Forest Lake, MN, says that whether his company buys molds for its OEM customers or the OEM customer buys the mold directly from the moldmaker, his company wants to be involved. "We're not moldmakers, but we provide engineering expertise. It takes a three-way joint effort to get the mold created, and we want to play a role." Plastech adds a nominal fee to the cost of the tooling to cover its engineering costs.

Mold buyers, particularly those at large OEM companies that might be distanced from manufacturing, sometimes don't realize that a mold isn't built in a vacuum. The mold is one component—a critical one—of a process. Everyone involved in that process needs to give input.

Having said that, lots of moldmakers prefer to work directly with OEMs because of certain advantages. Large OEMs are good prospects for upcoming products that will require molds. OEMs that do their own molding tend to invest in high-quality molds, since they will be running the production. Plus, there is less chance of something falling through the cracks if no middleman is in the picture. In some cases, payments are made in a more timely manner when dealing directly with the OEM. (See Chapter 9, "Paying the Moldmaker.")

FINDING THE PERFECT MOLDMAKER

As noted earlier, finding a moldmaker who meets your particular needs generally means visiting various shops and getting to know each one's capabilities. If you're an OEM engineer who knows plastics and understands molds, the search won't be difficult. However, if you're new to purchasing molds, you may find the size and fragmented nature of the industry daunting. Just remember, choose a moldmaker who can do the job you need done, with good-quality workmanship and on-time delivery, and at a fair price.

Since not all mold shops are created equal, here are a few things to consider.

Do the capabilities of the shop match the requirements of the mold I need built? For example, some shops specialize in large, multicavity tooling; some in unscrewing cap and closure molds; some in single-cavity prototype molds; some in large, single-cavity molds; some in very small, micro molds.

Over the years each has developed expertise in certain types of molds or has done a lot of work for specific industries. Some are expert at making molds for cellular telephone housings, or molds for business equipment, or molds for the medical industry. The expertise these shops have developed can be valuable to you and your project. These moldmakers understand the nuances of

Choosing a Moldmaker

molding certain types of parts, and exactly what is required of a mold.

Select a shop with the expertise and capabilities you need for your project.

What is this shop's record for on-time delivery? When you are facing a narrow time-to-market window, the last thing you need is a mold that's going to be weeks late. Obviously, some glitches will pop up along the way, but generally a good shop knows its schedule and, barring any major changes on your part, will be able to deliver within the time frame specified. Ask for references, and then check them out to confirm the company's track record for on-time delivery.

What is this shop's reputation within the industry? It's a small world in the moldmaking industry. Most shops know their competitors and which shops are considered tops in an area. Ask around. You can also call an industry association for a listing of member companies, which are held to a standard of ethics in business and quality in product.

Go through company brochures and look at the list of facilities (equipment, capabilities, and so forth) and read about the company, and then choose three or four that seem to fit your requirements. A personal visit is always recommended. Get to know the shop's management; walk around and see the facility for yourself. Ask questions.

It's not uncommon for OEMs to qualify their suppliers prior to releasing work to them for the first time. Maybe you have a Vendor Qualification survey form you could take with you to have the moldmaker fill out to become a qualified supplier. Most OEMs or molders want to be certain the moldmaker is on solid financial footing, has a good reputation for quality and on-time delivery, and has the capabilities and expertise to do the job.

A good source of information is *Customer/Supplier Evaluations: The Development of Standards for the Objective Measurement of Both* by consultant Bill Tobin. It contains good, practical advice, along with sample evaluation forms (also included with the book on disk) that protect all parties involved (see Table 5.1).

Business-savvy moldmakers might have a form for you to fill out as well, particularly if they have not done business with you previously. They may want references to determine the financial condition of your company, particularly if it is a small, privately held entity. (Moldmakers would be wise to use Tobin's book to help them choose their customers!)

All this checking each other out is just good business and part of building a relationship. Starting out on the right foot helps avert the potential for real disaster that can occur when the parties know each other only superficially.

THE TICKETMASTER SUCCESS STORY

When management at TicketMaster Inc., one of the largest sellers of event tickets in the United States, decided to develop a computerized, handheld, ticket-reading device, they realized they would have to get involved with the plastics industry. Because this was their first foray into "hardware," they were also aware that they were entering an unknown realm.

Table 5.1 Supplier Evaluation/Information Questionnaire (*Source:* WJT Assoc.)

Organizational Information

Company name:

Mailing address:

Shipping address:

Telephone number: (   ) _____

Fax number: (   ) _____

E-mail address: _____   System: _____

Subsidiary:

Company name:

Mailing address:

Shipping address:

Telephone number: (   ) _____

Fax number: (   ) _____

E-mail address: _____   System: _____

Nature of Business:

(Describe what kind of business you are in; i.e., plastic injection molding with full decorating and assembly capability.)

Table 5.1 Supplier Evaluation/Information Questionnaire *(Continued)* (*Source:* WJT Assoc.)

Corporate Organization:
Structure: (Sub-S Corp., Ltd. Partnership, Partnership, etc.)

Year started: (When did you open as a business?)

Officers: (List president and ceo, and how to contact.)

All other corporate disclosures are optional.

Financial:
(Name, address, and phone/fax numbers of merchant banker who can give a report on the company's financial integrity. Note: Make sure you know what this person will say and what financial reports are available ahead of time before giving blanket permission to your customers to talk with him/her. Also provide a Dun & Bradstreet or similar credit rating.)

Business Acumen:
Management discourse on operating philosophy. (Describe how inquiries are handled.)

Organization:
Provide an organizational chart with positions, and/or the names of people who hold these positions.

Excerpted with permission from Bill Tobin, *Customer/Supplier Evaluations: The Development of Standards for the Objective Measurement of Both* (1994).

The company hired a designer who was knowledgeable in plastic part design, an excellent way to start. Then management began seeking out moldmakers and molders for its project. Because the company was located in Tempe, AZ, they wanted a local moldmaker and molder. They didn't feel confident enough to choose a moldmaker on the East Coast or offshore, and thought that a local moldmaker could give them the level of comfort they needed as they embarked on this new project involving plastics.

Using the local Yellow Pages, the purchasing manager looked up several mold shops in the area, and, when the part prints were completed, sent out RFQs to several shops sight-unseen. Naturally, when he received the quotes back, numbers ranged all over the map.

One moldmaker, however, contacted TicketMaster to prequalify the RFQ and to determine more exactly what was wanted. When the tooling engineer at Mastercraft Mold discovered this was TicketMaster's first venture into the wonderful world of plastics, he immediately arranged a meeting in which company management could visit Mastercraft Mold and its sister company, molder Polycraft Industries.

Not only did Mastercraft personnel give TicketMaster's management a tour, they showed them the entire process from mold design through to completed product, including secondary operations and assembly. TicketMaster's management was amazed that there were so many steps involved in getting the five plastic components they needed to comprise their new product.

Beware the Low Bid

A moldmaker who serves as an expert witness in litigation involving so-called "bad" molds explains that trouble often arises because too many purchasers buy molds "based on price and ease of buying." In other words, they simply fire off RFQs to half a dozen shops, then pick the lowest bid.

This consultant stresses the importance of getting to know a moldmaker. Buyers need to "go to the different moldmakers and choose the best one for the job," he says. "Sometimes all they worry about is their budget, so they give the job to someone who doesn't have the capability to build the mold."

A mold purchaser must look at more than price.

An irrigation products company needed a set of unscrewing molds built for sprinkler parts. The company awarded the job to a moldmaking/molding firm that had been calling on it for a few months, inquiring about possible work, even though the moldmaker had no expertise in building unscrewing molds containing threads.

The irrigation company liked the low price the moldmaker bid on the job, which, in this case, reflected the moldmaker's lack of knowledge. As can be expected, the project turned into a disaster for both the moldmaker and the irrigation company.

The moldmaker's inexperience caused the molds to be completed three months late. Startup was a nightmare due to the many problems with the molds, and the customer refused to make the final payment on the molds. Everyone lost.

Having your molds built by a moldmaker with expertise in the type of molds you require is essential to success.

Mastercraft quoted and won the project, after Ticket-Master determined that the sister companies had all the necessary capabilities and expertise to design, build, and produce its new product. The fit was excellent for Mastercraft as well, since the company had expertise in designing and building molds for small electronic products, as did Polycraft in molding and assembling them.

As a result of a little initiative, careful planning, and taking the time to get to know each other, the project went smoothly. That's the advantage of prequalifying to all involved.

WORDS OF WISDOM

Unfortunately, there is no shortage of mold shops that will tell buyers whatever they want to hear just to get the job. One mold shop used to advertise, "Any Mold—Any Price." Others promise delivery dates that they know they can't meet, or agree to build a mold they know they don't have the expertise or capability to build.

It is common knowledge that once a moldmaker gets halfway into the mold building process, the mold purchaser won't—in most cases—pull the work because of the difficulty of finding another shop willing to take on a botched project or that can complete the job by deadline.

In the few cases where an incomplete or unsatisfactory mold was pulled from the moldmaker and completed at another shop, the mold ended up costing much more. Plus, the time lost in getting the product to market can be calculated in the thousands or even hundreds of thousands of dollars. It can't be stressed enough—prequalify and choose your moldmaker carefully.

> **Price, Delivery**
>
> Purchasers, particularly those who have dealt with the moldmaking industry for many years, say they need moldmakers to be honest with them about price and delivery. These, they say, are the two most critical elements of a mold build. They don't mind paying for design changes and other components that might have been missed, but they want the moldmaker to stick as closely as possible to the quoted price and delivery time.

HONESTY IS THE BEST POLICY

You've gone to a lot of trouble to find the right moldmaker for the job, so take the time to develop an honest, working relationship that will nurture a long-term, mutually beneficial partnership.

Always be upfront with your moldmaker about issues affecting the mold build. If you expect your moldmaker to be honest with you, you must also be honest with your moldmaker.

CHAPTER 6
BUYING MOLDS OFFSHORE—IS IT RIGHT FOR YOU?

Molds built offshore—in Taiwan, Hong Kong, China, and Portugal—came on the moldmaking scene in the early 1980s. They came with promises of half the cost and half the time it took to build molds in the U.S. For buyers, the prospect was intriguing, to say the least.

But the promises came with a downside. The molds coming from offshore often had substandard components and were made of substandard materials. Short lead times didn't take into account the several weeks on a boat it took the molds to get here. In some cases, molds arrived in poor condition, needing lots of rework prior to being hung in a press. Worst-case scenarios brought molds that were little more than expensive boat anchors when they landed.

In the 1990s, the molds improved. Moldmakers in Hong Kong and Taiwan caught on to the standards desired by U.S. companies. They can now build molds using the same components and steels as U.S. moldmakers, and they are eager to capture the majority of the world's mold building.

U.S. moldmakers bristle at the thought. They continue to believe that the best molds are produced in the United States by skilled craftsmen and craftswomen, a viewpoint that has a lot of support from the OEM community. In today's competitive environment, however, some mold purchasers look to buy molds from offshore sources that continue to offer lower-priced molds.

WHERE IS OFFSHORE?

When the subject of offshore molds comes up, lots of people think of Taiwan, China, and other Asian countries. However, Portugal has become a popular place to purchase offshore tooling, particularly for OEMs on the East Coast.

A trade magazine's coverage of Portugal's moldmaking industry[1] noted that, although some cost advantages exist in the relatively low labor costs of Portugal, when compared to Europe, at 10 to 15 percent, labor is a small and decreasing component of producing a mold.

Portugal is capable of building increasingly sophisticated tooling as a result of the heavy investment companies there have made in moldmaking technology. And, realizing that U.S. companies want high-quality molds, Portugal has responded by producing complex, expensive molds.

The same holds true in Taiwan. In the 1980s, the quality of molds coming from Taiwan was anyone's guess. Sometimes you got what you wanted; other times you didn't get anything close to what you had ordered. U.S. standards were an apparent mystery to Taiwanese mold shops.

Today, Taiwanese moldmakers can produce molds to U.S. standards using U.S. steel and components. That still doesn't mean that you don't have to be careful when you decide to have a mold built offshore, nor does it make it any easier to select an offshore moldmaker.

THE LOCAL ADVANTAGE

If you think finding a moldmaker in the United States that meets all your criteria is tough, try finding a moldmaker halfway around the world! So consider first the advantages to buying molds closer to home.

Local vendors speak your language, and they're close by. According to Roland Krevitt, a tooling engineer who has been purchasing tooling for major OEMs since the mid–1980s, "The language difference and the long distance away are major complications. To overcome the distance problems, someone like a tooling engineer virtually has to live in the vicinity of the tool shop to answer questions, monitor progress, and make sure the molds are built the way they are supposed to be built."

There's a lot to be said for having your moldmaker within easy traveling distance rather than half a world away.

Molds are built to U.S. standards. You want molds built using tool steel that is certified and with components that meet U.S. standards. Offshore moldmaking companies claim that they build to U.S. standards, but who can be certain? One of the biggest problems that still plagues purchasers of offshore molds is the use of

> ### What's This Made of?
>
> An OEM of disposable hospital supplies had a large mold made in Portugal. The core cracked the first day it ran; moldmakers tried to weld it, but could not because it was of a strange material they'd never seen before. They resorted to calling in a metallurgist to assess the metal, but to no avail. Ultimately, the company had to buy a new mold, since attempts to weld the crack permanently were for naught.
>
> This is just one of the many problems you can run into with an offshore mold when you don't specify exactly the type of steel or components you require, or when you don't watch over the mold carefully during the build by being on the scene.

mold steel that cannot be welded or that does not hold up under normal molding conditions.

Recourse is available. A mold built in the United States generally comes with some kind of guarantee to perform as agreed upon, or the moldmaker will make it right. An offshore mold is seldom shipped back to the country of origin for repairs or engineering changes, and U.S. moldmakers do not like to repair offshore molds because of the unusual problems they run into.

Although local shops may be willing to take on Engineering Change Orders (ECOs) and repair offshore molds (often with the hope that they will eventually get a mold build job from you), how long a local shop will be willing to support your offshore molds is a question you need to ask. Strategic alliances have been formed between U.S. companies and offshore moldmakers, and if an offshore mold is absolutely what you're going to buy, some of these companies will provide engineering supervision of your tool until it is delivered. They can also provide any engineering changes or tweaking the mold might need after it arrives in your or your molder's plant.

Cost savings quickly dissolve. You will save about 20 percent to 30 percent on the price of the mold, according to buyers of offshore molds. Once the mold is finished, "there will be several tryouts and changes that quickly eat up any savings," says Kravitt. "Molds built by offshore shops seem to require more tryouts to get them right than those fabricated by U.S. shops. If a tooling engineer is not onsite continuously, it is mandatory that he visit frequently. Trip costs can rapidly erode any hopes for savings."

Lead-time savings quickly dissolve. Your mold may be built two weeks faster than one made by a U.S. moldmaker, but if you ship it by transoceanic container freight, you can lose that time. Air freight is always available, but at a premium cost, which also cuts into any real savings.

OFFSHORE MOLD BROKERS

Mold brokers take your mold requirement or RFQ and shop, or broker, it around until they find the lowest bid in China or Portugal or Singapore. Then they may do little more than sit back and wait for the results.

Anyone who has ever had experience with offshore mold builds advises that a physical presence is almost mandatory to ensure you get the mold you want. Brokers who do not invest the time to visit the chosen shop to monitor the mold's progress usually put you on the wrong end of a bad deal. Hundreds of horror stories about mold brokers who fail to deliver on their promises kick around the industry.

If you decide to use a mold broker rather than oversee the offshore build yourself, be certain to enlist the aid of a mold engineering firm. You can find several good companies and individuals in the United States who are experienced tooling engineers, who know the moldmaking business, and who understand the intricacies of dealing with offshore mold shops. They guarantee that your mold will be what you want, because they make frequent visits to the mold shop and personally oversee the mold build.

SPECIFYING OFFSHORE MOLDS

If supplying U.S. moldmakers with exacting details of the mold you need constructed is essential, when dealing with offshore moldmakers it assumes even more importance.

Foreign shops can buy mold components from U.S. suppliers or can supply the equivalent according to U.S. standards, including the use of standard inch-feet measurements vs. metric units. However, you must specify exactly the mold steel and types of mold components you want, or you will get whatever the mold shop has available or what it "thinks" you want.

IS AN OFFSHORE MOLD RIGHT FOR YOU?

After you have evaluated the type of mold you need, then you can weigh the pros and cons of buying from an offshore source.

> **Quoted vs. Actual Cost**
>
> U.S. moldmakers often accuse mold buyers of failing to figure in the total cost of the mold, costs exclusive of the actual design and build of the mold. These costs include travel time and expenses ($5000 is the average cost of an airline ticket to Taiwan); time of the engineer who is away from his/her office for several weeks; hotels and meals, cab fares, and other miscellaneous costs. Also, the cost of shipping the mold can be a considerable expense when it's coming from halfway around the world.
>
> All these add up, making some mold programs more costly than if the mold was built in the United States.

Floyd Binder, senior tooling engineer for Teledyne Water Pik in Fort Collins, CO, uses both offshore (specifically Asian) sources and U.S. sources for his mold requirements. The company has its own in-house molding operations. Binder comments that over the past decade Asian moldmaking shops have improved in the area of quality.

"The ability of Pacific Rim shops has escalated to match the U.S.," he says. "Sometimes their equipment is as new or even newer than in U.S. shops. A majority of the Pacific Rim shops we visit have e-mail capability, so transfer of information isn't too difficult. Still, you have to deal with language barriers and time-zone delays."

Plus, there are always additional hidden costs. Moldmakers often say they see a "disconnect" among mold purchasers between the price of the mold and what the mold actually costs to bring to production readiness.

Therefore, you need to evaluate carefully what type of mold you're having built in Asia and why you want it built there. A small one-cavity mold that costs $25,000 in the United States might come in from an Asian shop with a bid of $20,000. Yet, by the time you factor in additional expenses, you might end up spending more than $30,000.

"We want to know what we are spending after we've brought the mold in-house. What does it take to bring this mold to our production level?" says Binder. "Our latest experience with Portugal resulted in a mold that ran well when it hit our press, but a lot of work went into getting it that way. We're putting a lot more effort into tracking the so-called hidden costs to see what the price of a[n offshore] tool really is."

Another tooling engineer for a major lawn-and-garden equipment OEM notes that he spends an additional $4000 to $5000 per mold getting it production-ready after it comes to the U.S. from China. He readily admits that management at his company only looks at the bottom line of the quotation when determining where to have molds built. The costs of bringing the molds to the U.S. and bringing them up to their molding production standards come from a separate budget.[2]

"It doesn't pay to go offshore unless you're putting out a lot of tools every year," says the consumer products tooling engineer. "You need large tooling packages and more than seven molds to make the expense of sending people to monitor the project worth it."

Mold buyers with offshore experience agree. The biggest cost savings is in larger, complicated tooling, they say, and labor is cheaper in Asia. Also, larger companies

Buying Molds Offshore—Is It Right for You?

that can afford to send people overseas to monitor the mold build are more apt to go offshore. Smaller companies without the manpower might use a broker or stay in the U.S.

Caveats about lead time exist as well. "We were hearing about fantastic delivery times, but we found they were mostly on soft tooling," the consulting engineer notes. "Because we do all hardened-steel tooling, mold development and build times are similar to that of U.S. moldmakers. The time savings is primarily in P–20 tooling, which averages eight weeks. With hardened tools, 14 to 16 weeks showed up on the quotes."

He also noted that the Pacific Rim shops he encounters are not very receptive to building hardened tooling. "Some shops are okay with it, but others don't even know what you're talking about."

Although this engineer has been forced by circumstances to purchase many of his molds offshore, he states, "We still believe you cannot buy molds anywhere in the world as good as those in the U.S."

Water Pik's Binder says that in his experience, and that of other tooling engineers in the business, molds built in a foreign country should be run in that country "under the jurisdiction of the local OEM or your local engineering office staffed with the proper disciplines, so that the responsibility is local and manageable and there is no language problem, rather than be shipped to the United States for molding parts here." Doing anything else guarantees problems, he maintains.

Notes

[1] See "Does Offshore Mean Budget Molds? Not Necessarily," *Injection Molding Magazine*, (June 1998), p. 90.

[2] Help in comparing the cost of a mold from offshore vs. U.S. sources is available in *Know the true Cost of Your Molds,* a booklet with a worksheet from the American Mold Builders Assn. To order, call (630) 980-7667.

CHAPTER 7
COMPUTERS, E-COMMERCE, AND MOLDMAKING

Electronic Data Interface, or EDI, is now an important element in moldmaking. Moldmakers across the country have Internet sites through which electronic transactions such as transferring data files to request a quote, to update part design files, or to track a mold build can be conducted.

Sipco Inc. in Meadville, PA makes it easy for customers to follow the progress of their mold's build—they simply log on to a secure site and "access their own progress 24 hours a day," says Kevin Maziarz, sales manager.

The company also has cut down on the time required to import CAD files, design molds, and generate CNC programs to cut steel. Using Pro/E, along with Pro/MoldDesign and Pro/Manufacturing from Parametric Technology (Waltham, MA), Sipco speeds up the entire design process significantly. A substantial percentage of the company's customers have the same system, so importing data files is error free.

This is a key point to keep in mind: Participating companies' systems must be identical or problems will most likely occur in the translation from one type of system to another (see "Estimating Mold Costs Using CAD," Chapter 4). A flyer, *Guidelines for Part Prints, Databases and Communication* is available from the American Mold Builders Assn.

CAD BARRIERS

In spite of all the computer-aided design, engineering, and manufacturing assistance, why are molds still built that can't run good parts? Anne Bernhardt, president of Plastics & Computer Inc., a software provider and consulting firm in Dallas, believes that CAD/CAE technology promotes errors because people have too much confidence that the computer systems will do everything—and they don't. "Management expects computers to reduce mistakes, improve quality, shorten lead times, provide documentation, and be a cornerstone to [concurrent] engineering," she says.

Plastic part designers can focus so completely on the part they're designing that they forget the mold and the molding. Designers who lack an understanding of molding often design in problems for the molder. Everyone involved needs to understand that part design and mold design are just two aspects of the entire process in which the goal is parts that conform to specifications and/or are functional.

Bernhardt believes the team approach is crucial to a project's success. She encourages companies to develop teams comprised of people from different disciplines and with varied experience levels.

Not only has CAD/CAE/CAM *not* eliminated the need for everyone from the designer to the moldmaker to the

Clean Database Files

Mistakes in data files are the most common cause of severe problems in the concurrent engineering process. According to one tooling engineer, "A database using solids modeling or a parent file must be established as gospel. All data must be contained in this single database." This makes clean database files an absolute must. Almost every file needs some degree of cleaning up after transfer from the OEM to the moldmaker. However, some are so cluttered that many hours- even days- are required. In fact, many moldmakers will not start the clock for the mold build until the data files are completely clean.

Moldmakers sometimes lament how much money they lose on projects because of cleaning up data files that were supposed to come to them clean in AutoCAD and Pro/E. If a file requires more than a minimal touch up, the customer should be notified and expect to pay an hourly rate for having the data cleaned.

David Johnson, a tooling engineer for Motorola in Ft. Lauderdale, FL, provided insight into what a clean database entails at Injection Molding Magazine's 2000 Management Conference held in Scottsdale, AZ. Clean, fully drafted part databases include fully defined

- Parting lines.
- Witness lines.
- Gate locations.
- Ejector pin locations.
- Markings (date code, recycle logo, vendor mark, part number).

Although most systems are capable of translating information from one format to another, crucial data can be lost due to the problems one system has in communicating with another. This is just a fact of life at this point in time. Therefore, care must be taken to ensure that all databases are sent as clean as absolutely possible.

molder to be involved in the program, it has also made it mandatory that all disciplines be involved when designing and building a mold. If not, not only will bigger mistakes be made, but they will also happen more rapidly and with greater negative consequences.

CONCURRENT ENGINEERING: FACT OR FICTION?

A lot of myth surrounds concurrent engineering—what it is and what it can do. It is *not* a cure-all for poor part design, glitches in the mold build, and missed lead times. It *is* a concept that can work for you if you use it correctly, and if you realize that it takes cooperation on both your part and that of your moldmaker.

Concurrent engineering means simultaneous part development, mold design, and mold build involving the key personnel of both the OEM and the mold supplier. Here are a few of the components to successful concurrently engineered projects:

Early involvement of your moldmaker. Without this, concurrent engineering cannot happen. The sooner you share the concept of the product with your moldmaker, the sooner he can begin to formulate in his mind which type of mold design is best for the part.

Tech Mold's Bill Kushmaul once said he is reminded of the old cigarette commercial (when cigarettes were still advertised) that said, "It's what's up front that counts." That, he said, should also be the moldmaker's slogan.

Good communication. The failure of projects to be completed on time, or projects that result in parts that do not conform to print or function to expectations, in many cases can be attributed to poor communication between the OEM and the mold designer and moldmaker.

Throughout the process of part development and mold design and build, each new idea and each change incorporated at the OEM level—whether in design or time frame—must be communicated through channels established at the outset of the project.

Compatible computer systems. Compatible hardware and software facilitate the concurrent engineering process. A computer-aided design system capable of running 3-D solids and surfaces and importing/exporting data files from and to other major CAD systems, plus a skilled systems manager or operator, should be in place with all involved companies.

Plastics & Computers' Anne Bernhardt notes that computer models are crucial to the entire process, because without computer models you can't ask the right questions. "Computer models help you catch the gotchas upfront."

Changes to data files. When making changes to data files, never assume the change you make was received and incorporated by the mold designer. Establish a system for confirming that the changes were received and incorporated.

Have an operating plan in place. You and your mold designer/moldmaker should develop an operating plan that is an integral part of the process from the onset, adhered to by every pertinent division of the company and, when applicable, by all independent contractors, including product designers, tool designers and builders, and molders.

As stated at the beginning of this chapter, concurrent engineering isn't a magic bullet to fix poor communications or lack of processes and procedures. When the guidelines are adhered to, and everyone plays by the same rules, it works extremely well for both moldmakers and their customers.

Several good books on concurrent engineering to help with implementing this process into your company's procedures are listed in the References section at the end of this book.

ONLINE BIDDING

Yes, it is possible to bid online for a mold. Although this type of bidding typically is conducted in regard to molded parts, some molds have been bid this way as well.

FreeMarkets On-line Inc. in Pittsburgh is one e-commerce company that offers OEMs a way to bid for work online in an auction-like situation. Online bidding is not especially popular with moldmakers or molders because they believe it's important for a buyer to really know the shop with which he chooses to do business. But the temptation definitely exists to go online and get a bid from a dozen companies without knowing which companies are really capable of building the mold.

Rarely do OEMs award work blindly to moldmakers or molders they've never surveyed, according to Glen T. Meakem, cofounder and CEO of FreeMarkets. He insists that, like offline quoting, the OEM prequalify the companies that participate in the bid process.

Computers, E-commerce, and Moldmaking

> **Don't Overlook Websites**
>
> The Internet is an excellent resource for learning about different moldmakers. Many have websites containing information about their companies, capabilities, personnel, and expertise.
>
> In addition, most industry trade organizations have put up websites listing their members' companies. AMBA (www.amba.org) provides information on specialities and lists moldmakers geographically. Hot links are also included.
>
> Moldmakers' websites often provide space for sending an RFQ or a request for more information. Note, however, that although a website can begin the process of locating a moldmaker, it is no substitute for visiting the shop before awarding a job.

Because of the custom nature of molds, and the extensive engineering evaluation and design work that OEMs require of their moldmakers, however, it's unlikely that online bidding will catch on in a big way. Mold buyers, particularly those from companies that buy molds on a regular basis, will continue to use tried-and-true suppliers.

The Internet enables concurrent engineering to be a reality, and creates a means of communication that results in ease of transfer of information. It does not guarantee that molds can be built trouble-free. The power of computers and the Internet is only as good the users' ability to check, recheck, and follow up.

CHAPTER 8: THE MOLD BUILD

The process of building a mold always includes glitches, questions that must be answered, and minor problems that need solving before the build can continue. This means ongoing communication is required between you and your moldmaker.

DELIVERY AS A CRUCIAL ISSUE

The date of delivery of a mold seems more crucial than even its price these days. In fact, customers have turned up the heat so much that today you can find molds being built in half the time it took a decade ago.

Much of this is thanks to mold design software programs that reduce some of the mundane work to a few key strokes. New technology, such as high-speed machining centers, CNC, and EDM, also helps reduce the number of hours it takes to build a mold.

Mold buyers are known to live or die by the date of promised delivery. They know that everyone from engineering and new product development to marketing and sales is awaiting a conforming part from the mold to proceed with manufacturing. Even entrepreneurs with

> **Planning Ahead**
>
> In today's build-it-now climate, customers are willing to pay premium prices for overtime in order to accommodate their aggressive timing requirements.
>
> If your program is under severe time constraints, consider what it is worth to you to support a compressed work schedule for the mold shop. In other words, build their overtime into the cost of the mold up front, so that the project can be completed in a realistic time frame. Don't wait until the last minute, and then hold the moldmaker's feet to the fire.

one product will promise retailers a date for the product's availability. For a variety of reasons, when a company doesn't have a promised product ready to ship when orders come in, disgruntled customers at best and lawsuits at worst can happen.

Yet, even the most punctual moldmaker can run up against an unforeseen glitch. For this reason, it becomes extremely important that both the buyer of the mold and the moldmaker communicate with each other about delivery requirements as well as the progress of the mold build, and that both honor their commitments.

One moldmaker suggests that two delivery dates per quote should be required:

- The mold design delivery date.
- The mold build delivery date.

The combined time that both take would produce the actual delivery date.

Moldmakers point out that when database issues or design changes by the OEM cause delays in the design process, the time to build the mold cannot be lessened to compensate. It doesn't work that way.

OVERTIME

So, how about a little overtime to meet a delivery date? Obviously, that's going to be costly, and those overtime hours are going to eat into profitability. Moldmakers have even been known to lose money on jobs when unforeseen problems—sometimes brought on by the customer, and sometimes by themselves—caused delays.

A member of the American Mold Builders Assn. wrote to the group's executive director, "One of the most costly factors of mold building is overtime. We, as moldbuilders, have always taken pride not only in our workmanship but also in meeting our delivery commitments. Our industry is unique in the fact that based on a quote, we will work overtime to meet a delivery deadline, in many cases at little or no cost to our customers. How did we get to this ridiculous point? Why paint ourselves into a corner by promising a delivery or accepting the almost always short lead times demanded by our customers?"

Why? Because the moldmaker traditionally has accepted the responsibility to deliver the mold when promised.

Karl Van Blankenburg, a moldmaker and president of the Southeastern Michigan Chapter of the American Mold Builders Assn., expresses the desire of most moldmakers to meet their customers' delivery schedules. "Work overtime as deemed necessary to meet your commitments and to accommodate the overcapacity workload when applicable," he says. "The cost of meeting those commitments far outweighs the cost of a dissatisfied or lost customer."

WHOSE FAULT WAS THAT?

Part design changes are inevitable. Rarely is an engineering data file cast in stone. However, you can't expect to make lots of engineering changes and still hold your moldmaker to the originally quoted due date or price.

Some changes don't affect the delivery date, but others can set it back a week or more. It depends on what areas of the mold are being changed, and how far along the moldmaker is in the build.

Moldmaking industry standards dictate that when a customer calls with an engineering change to the part, the moldmaker tells the customer by phone or e-mail that the change might affect the lead time and/or the price of the mold. Then, after a review of the change, the moldmaker can make the necessary adjustments to the schedule and projected completion date—and to the pricing if the changes warrant such an adjustment—and notify the customer in writing of these adjustments.

Some moldmakers will stop work on a mold build until all changes are approved and the effect on the mold's price and delivery has been determined. This practice stems from customers who make major changes to a part design that result in changes to the mold, yet still hold the moldmaker to the original price and delivery. Then there are the customers who tell the moldmaker, "We'll finalize design as we go along." This should not be mistaken for concurrent engineering.

Changes that require recutting steel or rebuilding electrodes often mean major delays, and the moldmaker is put on hold until these changes are approved by the customer. When customers drag their feet in approving design, they risk delaying delivery of the mold.

WHEN THE MOLD IS LATE

Unforeseeable events can cause a mold not to be finished on time. Yet, when it becomes apparent that the original date isn't going to be met, some moldmakers will not approach their customers with a new delivery date. Usually this is because contracts contain penalty clauses for late delivery. Mold purchasers know that moldmakers will give a delivery date with a short lead time just to get the job, and that a possible penalty is one way to ensure on-time delivery.

To avoid penalties, mold shops will pull out all the stops and work hours of overtime to get the mold completed, even if the delay was caused by the customer. Yet, as noted above, overtime is costly.

When should penalties be imposed? Most moldmakers believe that when the late delivery was caused by the moldmaker, a penalty should be imposed. However, if the customer was responsible, no penalty should be placed on the moldmaker.

WHEN IS A MOLD COMPLETE?

Most agreements say that final payment is due when the mold is complete. But when is a mold complete? When

> **Progress Reports**
>
> Tracking the progress of the mold is critical to ensuring that the various phases of design and build are on time. Progress reports are offered as a normal course of business by most mold shops. The reports can be in the form of a Gantt chart or a written report. Some moldmakers even offer e-mailed digital photos of the mold in progress. Be sure to ask your moldmaker for progress reports to keep you updated and to alert you to any glitches.
>
> Consultant Bill Tobin, WJT Assoc., Boulder, CO, offers a computer program called ToolTrack (see References), which runs an electronic spreadsheet and breaks down components of mold construction, providing weekly updates on the progress of the tool build.

the steel is all cut and put together? When the mold produces a part that conforms to print? When the mold is put into production?

"The customer does not understand that delivery dates are not the same as due dates," complains one moldmaker. "The completion timing on molds should include a tolerance of plus or minus a week or so, in case Joe messes up or decides to take a week off."

AMBA's Karl Van Blankenburg answers, saying that, semantics aside, "The customer perceives the timing you quoted to be the date when the mold is done, complete, ready for production."

As a mold purchaser, you have the right to expect that the mold will be finished on the date specified, since you are relying on the expertise of the moldmaker to meet your needs.

Industry standards say that a mold is complete when it produces a part that conforms to the final approved part design to which the mold was built. But when a customer continues to make revisions to the part design after the mold has run samples, who should absorb the cost? Although the parts conform to the specifications of the print, the changes are not part of the agreed-upon mold build. Rather, they are engineering changes initiated by the customer.

For example, a customer decides after the sampling that there needs to be more radius on one corner to make a mating part fit better. Should he hold up his final payment until this change is done? No! Engineering changes made to the part design after the fact are not part of the original contract. The buyer should expect to be billed for the added costs.

MAINTENANCE AND REPAIRS

Most contracts contain a clause regarding the routine maintenance and repair of your mold, and who bears the associated costs. Most molding plants have in-house mold maintenance and repair capabilities, even if they do not design and build molds.

Repairs that can be done in-house and are minor in nature, such as replacing an ejector pin or a minor weld, are usually absorbed by the molder, who builds the cost into the piece part price. Whether or not the repair is being done because the molder caused harm to the mold during the processing, or because of the age and general condition of the mold, is also a consideration. If the molder breaks the mold somehow during production, the molder usually covers the cost of the repair.

However, if the mold is nearing its end-of-life, has severe wear, and can no longer make a conforming part, major repairs by the moldmaker will be billed to the mold's owner. When purchasing a mold, you should discuss these matters with your moldmaker and molder to ensure that maintenance and repairs, as well as end-of-life issues, are understood by all.

A successful mold build doesn't begin when the moldmaker cuts steel. It begins at the initial concept of the component or part to be molded. A successful mold build is the product of good communication, good design (part design and mold design), and an understanding of what it takes from both the customer and the moldmaker to get a good end result.

CHAPTER 9
PAYING THE MOLDMAKER

The relationship between price and value is the most important issue customers need to understand, a moldmaker once stated. "The value of the mold to an OEM's new product cannot be underestimated, yet too many people see the whole moldmaking process as a royal pain," he maintained.

Buyers of molds have been known to dislike having to deal with moldmakers, primarily because buyers have trouble seeing the value the moldmaker brings to the table. However, the price of the mold reflects much more than the cost of steel or a machinist's hours or the electricity to operate the equipment. The price of a mold reflects the moldmaker's intrinsic value, which consists not only of the hard cost of operating a moldmaking business, but the intangibles of expertise, knowledge, talent, and creativity.

Then there's the value of the mold to the viability of the parts to be molded. Without a mold designed and built to provide optimum manufacturability, maximum levels of productivity or efficiency cannot be achieved.

While you must expect to pay for that value, room for negotiation is usually there. So what can you do if you like a moldmaker and want his shop to do your work, but the price is too high?

Just as there are varying classes of automobiles, there are varying classes of molds to meet specific requirements. Contrary to what some buyers think, not every mold for every customer needs to be a so-called "Class-A" mold. Some "lesser quality" molds will do the job just fine, depending on factors such as number of parts needed annually, the type of material being run, and the length of the program.

Some mold shops specialize in high-end molds and will not build anything else for their customers. This is their niche, and, consequently, their customers tend to be those companies that require that type of mold and are willing and able to pay for it. Other shops specialize in prototype molds, aluminum molds, or molds that will provide conforming parts but at less cost.

If you get a quote from a moldmaker that seems high but you like the quality of the shop's work, go to the moldmaker and discuss the situation. See if there are alternatives he or she will consider. Sometimes that price might be the best he can do, but in other circumstances, ways to reduce the cost and still get you the quality you need are there.

PAYING FOR THE MOLD

When it comes to terms of payment, options vary. However, most mold shops adhere to the widely accepted industry standards that follow. These standards typically involve an initial down payment that accompanies the purchase order, then payments that progress throughout the course of the mold build, and a final payment made upon completion of the mold or approval of first sample parts, depending on your agreement.

Most mold shops will be creative and find a way for payments to be convenient for the customer while keeping their own cash flow in good order. This is why they are adamant about receiving payment in accordance with the terms they establish.

Here are payment terms that are fairly standard in the industry:

1. The down payment is one-third to one-half of the mold's total cost, although some shops require only one-quarter down. This payment typically comes with the purchase order for the mold, and provides the go-ahead for the moldmaker to purchase steel and mold components, as well as to begin designing the mold.

2. Progress payments are made throughout the course of the mold build. Usually, these payments are of some predetermined percentage of the total cost, and made at specified intervals. For example, if the total cost of a mold is $90,000, a moldmaker might require a one-third down payment of $30,000 to begin the project. If the project is scheduled for a 16-week delivery, a progress payment of another one-third, or $30,000, might be made at the halfway point.

3. At the end of 16 weeks, another payment of $30,000 would be due upon completion of the mold.

> **Paying for an Offshore Mold**
>
> Most foreign companies will accept a letter of credit from any major U.S. bank. Some will accept your business credit card. If you work for a major global OEM, most likely you will have no problems getting foreign suppliers to work with you on terms suitable for both companies.
>
> If you work with a mold broker, generally you pay the broker and he pays the moldmaker.

Variations of this formula accommodate varying needs of customers. Some moldmakers take 50 percent down, 30 percent at the halfway point, 10 percent upon completion, and another 10 percent upon final part approval.

Most mold shops' payment policies are equitable, and most will work with you to establish terms that you both can live with. However, with today's moldmakers becoming more astute in terms of business, they also are becoming increasingly strict with regard to payment. A few have even stopped work on a mold when payment was not received within a reasonable amount of time—yet another reason why communication between you and your moldmaker is critical to the overall success of your project.

AMORTIZING

When the moldmaker and the molder are the same vendor, amortizing the cost of the tooling into the cost of the parts can be possible. This is becoming more common as costs-to-manufacture rise, as it allows the mold buyer to reduce the project's initial outlay and spread the expense evenly throughout the life of the program.

The downside is that it forces the molder to finance the mold. As the moldmaker, the molder has already covered the costs associated with designing and building the mold and must now count on the volume of parts to pay for the mold as well as the costs associated with molding the parts.

Some large molding/moldmaking companies are adopting amortization for their major customers. John Weeks, president of Precise Technology Inc. in North Versailles, PA, notes that from 1991 to 1998, the company amortized $980,000 worth of customer assets (tooling, molds). In 1999 to 2000, the company amortized or expected to amortize $3.8 million in customer assets.

That's a significant amount of money for a molder to put on the line for a customer, and it usually requires a commitment on the part of the customer to ensure that the tooling is paid for in the event the program does not yield the production quantities anticipated.

Consultant Bill Tobin advises that when a moldmaker/molder is asked to amortize the cost of the mold, he should examine the cost and the life of the product to be molded. Figure the total amount of principal and interest as if the mold were being financed on a credit card, advises Tobin. Weight the costs so that the interest is paid first and the tooling cost is paid at the end of the contract.

Tobin also stipulates that the molder be the exclusive vendor for the part or product; that a schedule of repayment (ownership of the tooling) be available on request; and that a lien on the mold be included until the mold is paid for in full. Some states now have mold lien laws that allow a molder or moldmaker to hold tooling until paid for.

> **Discuss Options**
>
> Most moldmakers want every mold to be a work of art- that's a fact. This has to do with a moldmaker's pride in his trade, in his creative abilities, and in his skills. Sometimes such pride stands in the way of providing customers with an optimum mold that fits their needs at a cost-effective price.
>
> Discuss what you really need with the moldmaker of your choice. See if there are alternatives he or she will consider.

"At the end of the projected amortization time, which is a date and not a volume of parts molded," says Tobin, "if the amount of product purchased hasn't fully paid for the tooling, the client agrees to pay in one lump sum the outstanding balance of the amortized amount."[1]

WHEN THE MOLDER BUYS THE MOLD

When you give your purchase order to a molder, and he or she in turn contracts with a moldmaker, your down payment and progress payments are subsequently sent to that molder. The molder then pays the moldmaker. Or he's supposed to. If he doesn't, the moldmaker may just stop work and delay your project.

At this point, you may need to turn up the heat on the molder. Molders do not like to look bad to their customers, especially major OEM customers, and applying a little pressure to get him to pay the moldmaker usually is effective.

Notes

[1] See "The Business of Molding #20: Tooling Amortization—How Not To Get Burned," *Injection Molding Magazine*, (January 1998), p.73.

CHAPTER 10
GUARANTEES AND LEGAL CONSIDERATIONS

When two companies get together, there will always be legal considerations. The moldmaking industry is no exception. Moldmakers can get into disputes with their customers over a mold and wind up in court. In such cases, a winner seldom emerges.

It can be said with great accuracy that quite a few moldmakers have just a little knowledge when it comes to the legalities of contracts. For years, moldmakers and their customers did business on a handshake. A mold buyer could phone and tell the moldmaker to add a radius to this corner or reduce the wall section, and it was done. Never was there a question that the moldmaker wouldn't get paid or that the buyer would turn around and say he never made such a change.

Today, however, most moldmakers want contractual agreements with their customers. Yet their lack of sophistication in developing contracts fair to both parties can prevent them from going to that step in the relationship.

In truth, a contract doesn't have to be complicated. A purchase order in and of itself is a contract. So are acceptance forms. Quite simply, a good contractual agreement balances the risks and rewards for each party, as well as contains the essential terms of doing business together.

THE CONTRACT

Of all the players in the plastics industry, moldmakers are probably the least sophisticated and the most informal when it comes to doing business, according to industry attorney Mark Mahoney. He notes that this fact is often to their detriment, as well as to the detriment of their customers.

For years, moldmakers have done business on a handshake, yet they're doing business with customers such as Ford, GM, Hewlett-Packard, and IBM—large corporations that have deep legal pockets. In a legal action, the moldmaker usually finds he's no match for these companies' legal departments.

This is why Mahoney says the handshake days are over, and a moldmaker's business agreement must be formalized with a contract. Mahoney offers the following contractual tips[1]:

1. Maintain good documentation on every mold project. Given the number of times purchasing agents, engineers, and project managers shuffle within an OEM, "often the only thread that holds a moldmaking project together is the documentation that goes with it."

Mahoney also stresses that any and all of this documentation can be used in court if someone is sued.

2. Include provisions to protect both parties.
- The moldmaker should have interest on the debt provided for should the OEM fail to pay on time.
- The OEM should provide for penalties if the moldmaker is late due to his own fault.

3. Provide for arbitration and attorney fees.
Arbitration is less costly and time-consuming than a lawsuit.

4. Make reference to trade and industry standards.
If, as a purchasing agent, you are unsure of a standard, reference guides can provide the information.

5. State which entity owns the molds.
Generally, a contract will also contain language that declares the buyer's remedy is limited to refund of purchase price or replacement of defective parts, and disclaim consequential damages. This should not come as a surprise to the buyer.

WHAT GUARANTEES TO EXPECT

Although few moldmakers are willing to guarantee that molds will run for a specific number of cycles, guidelines from the Society of the Plastics Industry (see References) for molds built of the various types of materials should help you predict the life of the mold under normal conditions, as well as how many parts can be expected from

> **A Lifetime Guarantee?**
>
> Never make a decision on which moldmaker to give a mold job to based solely on cycle time estimates. A moldmaker can always give you an outrageous estimate just to get the work. This is where "cycle counters" are coming into play. These automatic counters on the mold keep track of the number of cycles a mold has completed so that maintenance and repair downtime can be calculated, and issues such as who pays for this can be addressed in the contract.
>
> There is a growing trend, particularly in the automotive industry, to make moldmakers guarantee their molds for the life of a program. Moldmakers are resistant to this pressure because they have little or no control over a mold once it leaves their shop. As a result, moldmakers are being advised not to guarantee a mold to run for more cycles than the standards established by the Society of the Plastics Industry.

certain types of molds. Most moldmakers will honor these SPI guidelines.

Guarantees on delivery and the mold are given by Accura Tool & Mold Inc., Crystal Lake, IL. The company began giving guarantees in 1995 and has found them beneficial to both itself and its customers. Accura evaluated the number of jobs it was late in delivering, what that meant for both the customer and for Accura, and concluded that missed deliveries caused headaches for everyone.

So it began offering a written delivery guarantee to a limited number of customers stating that if Accura fails to deliver on time, the company will impose a penalty on itself of .5 percent of the total quoted price per workday. It then decided that if it guaranteed delivery, it should also guarantee what it was delivering: "Critical molding dimensions certified in steel to customer-supplied shrinkage where quoted; certified steel and heat-treat on molding components; cavities fully interchangeable where quoted; and as-built drawings provided."

It will impose the same penalty as above on itself "in event of a return to Accura due solely to our mistake from receipt of the mold/inserts and necessary inspection information, to be offset by the number of workdays of early delivery."

Brian Beringer, general manager of the company, says one reason the company offers the guarantees is because customers will drag their feet and not make timely decisions. Having the guarantee in writing, including the completion dates of the various stages of the mold build, encourages timeliness on the part of the customer and makes for timely delivery of the finished mold. It places the responsibility on the customer to get the necessary information to Accura to complete the mold on time.

MOLD LIEN LAWS

As a mold purchaser, you should be aware that lien laws exist for the specific purpose of protecting the moldmaker/molder against nonpayment.

Nearly half of the 50 states have lien laws designed to protect primarily molders in cases where customers do not pay for molded products. However, moldmakers are also using the lien laws as a means to collect payment.

Because in most cases the mold leaves the moldmaker's premises for tryouts and production, he or she has little recourse except to sue if the final payment isn't made.

The mechanics lien law that exists in most states was used for years to include moldmakers and molders. However, some years ago the Society of the Plastics Industry deemed it necessary to protect these two groups more specifically. Thus was born the mold lien, or mold retention, law. Moldmakers in several states are currently attempting to implement amendments to their state's current mold lien law to include recourse for moldmakers who no longer have possession of the mold.

ALLEVIATING MISUNDERSTANDING

Contracts allocate risk by outlining each party's responsibilities before a crisis arises, rather than after. Allocating risk is not a matter of automatically shifting risk to the other party, but of establishing who can best bear the risk in specific areas.

For example, perhaps the buyer will agree to assume all the risk associated with product liability because the OEM has good product liability insurance.

Contracts put in writing outline a mutual understanding of each party's role in design, development, and production of the mold and molded parts. This helps alleviate misunderstanding as to what needs to happen and when.

Lastly, contracts provide predictability. Verbal agreements can lead to situations in which parties involved have vague or selective memories. When everything is in writing, there's no opportunity for one party to take advantage of the other based on what he or she thought

The Moldmaker's Point of View

As a buyer, you can expect moldmakers to include additional protection in contracts:

1. Delivery is conditional upon receipt of complete drawings, databases, specifications, source code (if necessary), and payment in accordance with the established terms.
2. No warranties that extend beyond [SPI standards] and no promises of a cycle time or life span for the mold (the moldmaker cannot predict manufacturing circumstances or environments).
3. Specify that all legal action be filed in the home county and state of the moldmaker.
4. A provision for arbitration and attorney fees. As noted, arbitration is less costly and time-consuming than a lawsuit.
5. Provision for a lien on the mold to secure payment. The lien can be specific to the tooling used to create the debt, or it can be general for all tools held in the moldmaker's possession.
6. Terms of payment spelled out and insistence on adherence to those terms.

(or said) was correct. Players or circumstances can change, but the contract remains the same.

If you are new to the position of buying molds, a good contract will help you understand the rules of dealing with a particular moldmaker. Also, ask about your company's history in dealing with its mold suppliers so that you have a good feel about what the moldmaker has contributed in the past, as well as what he can contribute in the future.

> **Change Authorization**
>
> If your moldmaker insists on a purchase order for each change, understand that he is only trying to protect both his interest and yours.
>
> Consider the case of a subsidiary of a large medical device OEM, which wanted a change to a tool that belonged to the parent company. The tooling engineer of the subsidiary called the change to the moldmaker. When asked about a purchase order, the engineer said a verbal agreement was all that was necessary.
>
> However, when the moldmaker invoiced the subsidiary, it was passed on to the parent company. The parent company, because they had not approved the change and had not issued a PO, refused to pay the bill, sticking the moldmaker with the cost.
>
> Only the person authorized to make changes should request the changes.

Notes

[1] See "10 Common Contract Problems Encountered by Moldmakers," *Injection Molding Magazine*, (August 1999), p. 71.

CHAPTER 11
A GOOD RELATIONSHIP FROM THE START

Buying a mold is a complex and sometimes tedious process, but a few rules can help make the process easier. They're not written in stone, but you'll find that to most moldmakers they are gospel.

1. Send an RFQ that is as detailed as you can make it. Don't make the moldmaker guess what you want. Be specific about the type of mold, the number of cavities, the steel, expectations of mold life, and any guarantees you'll need.

If you aren't certain about any of these items, get input to help you determine exactly what type of mold is best for your part requirements. The more detailed the RFQ, the more accurate the moldmaker's quote will be.

2. Be honest about why you are requesting a quote. If you need a ballpark figure to submit to marketing, say so. But don't ask for a complete engineering evaluation and quote, then casually mention it's just a preliminary quote on a project that's at least a year away. Quoting is time-consuming, and moldmakers want to spend their time quoting jobs that have good promise of becoming a reality soon.

3. Respect the intellectual property of the moldmaker.
The knowledge and creativity a moldmaker has acquired are his or her intellectual property. Keep those ideas and suggestions confidential.

4. Consider the benefits of forming a true partnership with your moldmaker(s).
Bring in him or her early on your project for input; work with him in regard to costing goals and budgets. The best purchasing is done by those who truly know their suppliers and play as a team with them to the benefit of both companies.

5. Communicate with and solicit communication from your moldmaker on a regular basis.
Many provide Gantt charts or other types of progress reports to let you know where the mold build stands. Request that information from the start.

6. Make your payments on time per the agreement.
Few moldmakers can afford to play banker. Building a mold entails expenses on their part.

7. Changes to the part design can mean changes to the mold
Remember, the more changes you make during the mold build, the less likely you are to get a mold in the lead time or at the price quoted.

8. Define up front when the mold is considered complete.
When it is shipped? When it is sampled? Usually a mold is complete when it is capable of producing a part according to the specifications of the print.

9. If it sounds too good to be true, it probably is.
You may find a moldmaker who quotes very low on a job. Maybe he's hungry, or maybe his overhead is low so he can price low. However, any quote that comes in too low might not be the bargain that it appears to be.

THE MOLD YOU NEED
When purchasing a mold, it's especially true that you get what you pay for. Your molded parts are only as good as the mold they come from, so be sure your mold is optimum to mold the parts you need for the life of the program.

GLOSSARY

Apprentice moldmaker. A person in training to become a moldmaker.

Automatic mold. A mold that repeatedly goes through a processing cycle, including ejection, without human assistance.

Backing plate. A plate used as a support for the cavity blocks, guide pins, and other components of a mold.

CAD/CAM. Computer-aided design and computer-aided machining, capabilities that most mold shops use to assist in designing and building molds.

CAE. Computer-aided engineering.

Cavity. The part of the mold in which the outside of a part is formed.

CNC. Computer Numerical Control, or controls found on machine tools used by moldmakers, programmed to perform cutting functions and tool changes automatically.

Cold runner system. A means to route melted plastic from the molding-machine nozzle through a sprue and down symmetrical, balanced channels equally to all parts in the mold being filled.

Concurrent engineering. The performance of part design and mold design/mold build simultaneously among all parties involved.

Conventional or standard mold. The most common type of mold used to produce injection molded parts. Also called a two-plate mold.

Cooling channels. Channels or pathways located within a mold base through which water or another cooling medium is circulated to control the temperature of the mold's surface.

Core. The part of the mold that forms the interior of a part. Also part of a complex mold for undercut parts. Cores are usually withdrawn to one side prior to the main sections of the mold opening.

Custom injection molder. A company specializing in the molding of parts to the specifications of another firm that handles the sale and distribution of the item, or incorporates the custom-molded component into one of its own products.

Cutter path. The path a drill bit or cutter tool follows in cutting and shaping the steel that forms the mold.

Cycle (molding). The time it takes to complete the sequence required to produce a plastic component, generally from closing of the mold to ejection of the part, or between a certain point in one cycle and the same point in the next.

ECO. Engineering change order. Usually issued by a customer's tooling engineer or purchasing agent when a change to the mold is needed.

EDM. Electric discharge machine. A machine tool used in the manufacture of electrodes that are in turn used to make the details of a mold's core.

Ejector pin. The rod, pin, or sleeve that pushes a molded part out of a mold as the mold opens. Also called knockout pin.

Glossary

Electronic data interface. Computer capabilities that allow the exchange of part prints, mold prints, and other information between a customer and a moldmaker. Also called electronic data transfer.

Family mold. A multicavity mold in which each cavity forms a different component that will be needed to assemble the finished product.

Fan gate. A shallow gate that becomes wider (and usually thinner) as it extends from the runner to the cavity.

Finish. The surface of a molded part.

Flash gate. A long, shallow rectangular gate in an injection mold, extending from a runner that lies parallel to an edge of a molded part along the flash or parting line of a mold.

Footprint. The amount of space a mold occupies.

Gate. The opening through which melted plastic enters the mold cavity.

Guided ejection. A mechanism to control alignment of the ejector pins and ejector retainer plate.

Hot manifold mold. A mold in which the portion of the mold containing the runner system has its own heating elements that keep the plastic resin in a state ready for injection into the cavities.

Hot runner mold. An injection mold for thermoplastics in which the runners are insulated from the chilled cavity plate so that they remain hot during the entire cycle. The plastic in the runners remains molten and is not ejected with the molded part.

Injection molding. The process by which objects are formed when a plastic material is fed through a nozzle into a mold.

Lead time. The time it takes to design and build a mold; often referred to as the estimated time of completion of a mold.

Machine tool. Equipment used in the production of molds, tools, jigs, and fixtures.

Mold. As a noun, the cavity, core, and base components that comprise the tool in which a plastic part is formed or molded. As a verb, to shape or form a plastic part.

Moldmaker. A person who has developed an expertise in the various machining, fitting, and assembly functions required to build a mold used in the production of plastic components.

Multicavity mold. A mold with two or more cores and cavities; a mold that produces more than one part per cycle.

Multiunit die. A mold in which several sets of cores and cavities can be used interchangeably in a common base. Master-Unit Die (MUD) is the tradename of the company that developed multiunit dies.

OEM. Original equipment manufacturer; a company that develops, markets, and sells its own branded products.

Parting line. The mark on a molded part indicating where the two halves of the mold met in closing prior to part formation.

Pinpoint gate. A restricted opening through which melted plastic flows into a mold cavity.

Progress payments. Payments made by the OEM customer to a moldmaker at specific intervals during the mold build.

Prototype mold. A temporary or experimental mold, often made from light metals such as aluminum in order to obtain information vital to the development to the final mold and/or part design.

Rapid tools. Molds made with alternative materials such as epoxy or composites that can be completed in a few days and used for R&D purposes.

Glossary

Request for quote. A formal request for a bid that contains the specifics of the mold to be built.

Runner. A feed channel in an injection mold that connects the sprue to the cavity gates. The term is also used for the plastic piece formed in this channel.

Selected laser sintering (SLS). A process of building a prototype part or prototype mold using fusible powders and a laser beam. Particles are fused or sintered together without melting the part.

Slide-core mold base. Generally used to mold a part when an undercut or coring feature is required that could not be formed and ejected through a standard mold opening.

Soft tooling. Tooling made from materials other than hardened steel, such as aluminum, P–20, epoxy, or RapidSteel materials.

Sprue. The opening in an injection mold through which material is fed into a runner system and subsequently into the part(s).

Sprue gate. A passageway through which melted plastic flows from the nozzle to the mold cavity.

Stack mold. Essentially, two molds stacked back-to-back sharing a common plate.

Stereolithography. Also known as SLA, and similar to SLS. Used for developing prototype parts.

Stripper plate. A plate that strips molded components from core pins, usually set into operation by the opening of the mold.

Submarine gate. Also called subgate or tunnel gate. A type of edge gate where the opening from the runner into the part is located below the parting line or mold surface, as opposed to conventional edge gating where the opening is machined into the surface of the mold or mold cavity.

Texture. Put on the cavity side of a mold when a particular surface finish is required on a part.

Thread-forming (unscrewing) mold. Used to form threads on a plastic component that automatically unscrews the part from the mold to maintain the integrity of the threads.

3-D solids modeling. A computer program that shows all the features of a part or mold in relation to each other.

Three-plate mold. A mold with three sections instead of two to allow for easier fill and for parts where cosmetics are critical.

Unassisted or unattended machining. Tasks performed automatically through computer programs without the aid of a moldmaker.

Waterlines. (*See* Cooling channels)

BIBLIOGRAPHY

Backhouse, Christopher J. and Naomi J. Brookes. *Concurrent Engineering: What's Working Where* (1997), 248 pp., $59.95.

Fleischer, Mitchell and Jeffrey K. Liker. *Concurrent Engineering Effectiveness: Integrating Product Development Across Organizations* (1997), 506 pp., figures, tables, exhibits, references, index, appendix, $49.95.

Graham, Len, and Tech Mold. *What Is a Mold? An Introduction to Plastic Injection Molding and Injection Mold Construction* (1998), 143 pp., illus., glossary, index, appendix, charts, $50.00.

Hatch, Bob. *On the Road With Bob Hatch: 100 Injection Molding Problems Solved by IMM's Troubleshooter* (1997), 185 pp., illus., tables, index, glossary, solution finder, $49.00.

Huthwaite, Bart. *Concurrent Engineering User's Guide* (1995), 90 pp., $35.00.

Mahoney, Mark. *Business Practices for Manufacturers: A Guide for Business and Commercial Contracts* (1994), 59 pp., disk with sample forms, glossary, $49.95.

Society of the Plastics Industry. *Customs and Practices of the Moldmaking Industry, AR-101,* rev. ed. (1996), 20 pp., $20.00.

Society of the Plastics Industry. *Mold Finish Guide, AR-106* (1998), 2 pp. plaque, $10.00.

Tobin, William J. *Injection Mold Tooling Standards: A Guide for Specifying, Purchasing and Qualifying Injection Molds* (1993), 95 pp., $20.00.

Tobin, William J. *Tool Track: A Spreadsheet-Based Progress Report for Predicting Delivery of Injection Molds* (1994), 1 disk, $50.00.

These books may be ordered from the IMM Book Club: by phone, (303) 321-2322; by fax, (303) 321-3552; by e-mail, dgolanty@immnet.com; or online at www.immbookclub.com.

INDEX

Note: An *f.* refers to a figure; a *t.* refers to a table; a *sb.* refers to a sidebar.

A
Aluminum 37
Aluminum molds 40
Amortizing tooling cost 116–118

B
Bargains 129
Bridge tooling 40
Buying molds 87–96

C
Change orders 108–109, 128
Clean database files 99*sb*.
Cold runner 26–27, 27*f*
Completion 129
Computer-aided design/
 computer-aided machining
 (CAD/CAM) 5
Computer numerical controlled
 (CNC) 4
 problems 98, 100
Concurrent engineering
 100–102
Contracts 120

Conventional molds 16, 17, 17*f.*
Conventional runner 26

D
Design for manufacturability
 (DFM) 16

E
Electric discharge machines
 (EDM) 4
Electronic data interface (EDI) 97
Expected service life 15

F
Family molds 22–24, 23*sb*

G
Guarantees 121–123
 a lifetime guarantee 122*sb*.

H
High-speed machining 44–45
Hot runner 27–30
 conversion comparison 29*t*.

I

Intellectual property 54–56, 128

L

Lead time 71–73, 72sb.
Legal considerations 119–126
   change authorization 126
   the moldmaker's point of view 125sb.
Lien laws 123–124
Local vendors
   advantages 89–91

M

Maintenance 39, 111–112
Manufacturability 16
Manufacturing cells 7
Molds
   considerations 14
   criteria 15
   features 11–13, 15
   types 16–26
   what's this made of? 90sb.
Mold brokers 91–92
Mold building 105–112
   completion 109–111
   delivery dates 105–107
   engineering changes 108–109
   late penalties 109
   overtime 107–108
Mold quote form 60–61
Moldmaker
   beware the low bid 84sb.
   capabilities 78–79
   choosing 75–86
   delivery time 79
   evaluation 78
   price and delivery 86sb.
   relationships with molders 76–78
   reputation 79–80
Moldmakers and design 39sb.
Moldmaking 3–4
   future 7–9
   new technology 4–7
   overview 3–4
Multiunit die 24

N

Nickel composite tooling 47–48

O

Offshore molds
   companies 8
   defined 88
   evaluation 93–96
   quoted vs. actual cost 93sb.
   specifying molds 92
Online bidding 102–103, 103sb.

P

P-20 steel molds 37, 40
Part design 15, 38
Partnership 128
Payment 113–118, 128
   molder to moldmaker 118
   offshore molds 116sb.
Personnel 6
Productivity 39
Program life 38
Progress reports 110sb.
Prototype molds 43–44
   choosing a method 49
   prototype mold, why 45sb.
Pull ahead tooling 40

Index

Q
Quotes
  calculation methods 67–68
  exceptions 64
  negotiating 71
  sample mold quote 69t.
  understanding 68–70
  what it takes 63sb.
Quoting process 52–54

R
RapidSteel 2.0 46
Relationships, building good 127–129
Repairs 111–112
Request for quote (RFQ) 11, 51, 127
Requesting a quote 51–60, 62–74
  quantity of requests 51–52
Resin properties 38
Responsibilities 124–125
Runner systems 26–30
Runnerless 27–30

S
Slide-core mold base 16, 18f.
SLS vs. Aluminum 45–47
Society of the Plastics Industry (SPI) 57
Soft tooling 37–38
Solid-model viewing packages 66
Specifying molds 11–42
  cost 40–41
  number of cavities 30–36
  soft tooling 37
  spares necessity 36
  texturing 37
Stack molds 24–25
Standard molds 16, 17, 17f.
Standard practices 54
Stereolithography (SLA) 44
Stripper-plate molds 18, 19f.
Supplier evaluation/information questionnaire 81t.

T
Texturing 37
Thermoset composite board 49
Thread-forming molds 20–22, 21f.
3-D solids modeling 66
Three-plate molds 19, 20f.
TicketMaster success story 80–85
Two-component moldS 25–26

U
U.S. standards 89

V
Volume 15

W
Websites 103
Workload, leveling the 70sb.